THE MIRACLE OF MAN

THE MIRACLE
OF MAN

THE FINE TUNING OF NATURE
FOR HUMAN EXISTENCE

MICHAEL DENTON

SEATTLE   DISCOVERY INSTITUTE PRESS   2022

Description

For years, leading scientists and science popularizers have insisted humans are nothing special in the cosmic scheme of things. In this important and provocative new book, renowned biologist Michael Denton argues otherwise. According to Denton, the cosmos is stunningly fit not just for cellular life, not just for carbon-based animal life, and not even just for air-breathing animals, but especially for bipedal, land-roving, technology-pursuing creatures of our general physiological design. In short, the cosmos is specifically fit for creatures like us. Drawing on discoveries from a myriad of scientific fields, Denton masterfully documents how contemporary science has revived humanity's special place in nature. "The human person as revealed by modern science is no contingent assemblage of elements, an irrelevant afterthought of cosmic evolution," Denton writes. "Rather, our destiny was inscribed in the light of stars and the properties of atoms since the beginning. Now we know that all nature sings the song of man. Our seeming exile from nature is over. We now know what the medieval scholars only believed, that the underlying rationality of nature is indeed 'manifest in human flesh.' And with this revelation the... delusion of humankind's irrelevance on the cosmic stage has been revoked."

Library Cataloging Data

The Miracle of Man: The Fine Tuning of Nature for Human Existence by Michael Denton

256 pages, 6 x 9 x 0.6 in. & 0.9 lb., 229 x 152 x 15 mm & 0.4 kg.

ISBN-13: Paperback: 978-1-63712-012-5, Kindle: 978-1-63712-014-9, EPub: 978-1-63712-013-2

Library of Congress Control Number: 2022935635

BISAC: SCI036000 SCIENCE/Life Sciences/Human Anatomy & Physiology

BISAC: SCI019000 SCIENCE/Earth Sciences/General

BISAC: SCI015000 SCIENCE/Space Science/Cosmology

BISAC: TEC056000 TECHNOLOGY & ENGINEERING/History

Publisher Information

Discovery Institute Press, 208 Columbia Street, Seattle, WA 98104

Internet: discoveryinstitutepress.com

Published in the United States of America on acid-free paper.

First edition, first printing, May 2022.

ADVANCE PRAISE

While there is a general awareness of the fine tuning of the various laws and constants of physics rendering our planetary home particularly well suited for intelligent life, Michael Denton describes an additional astonishing array of qualities demonstrating prior fitness for complex carbon-based, high-energy, metabolically efficient life that takes the fine tuning in a different direction and to an exceptional degree. He cleverly describes the amazing fitness of oxygen, nitrogen, and water in both the hydrological cycle and in the respiratory and circulatory system. He highlights some surprising and intriguing observations, such as the relationship between the tension in small blood vessels and their ability to withstand relatively high hydrostatic pressures courtesy of the counterintuitive characteristics of the law of Laplace. Denton describes not just amazing and specific adaptations but the surprising prior fitness of basic physics and chemistry, a peculiar challenge to any naturalistic explanation and reminiscent of remarkable foresight. Teleology is evident everywhere you look.

—**David Galloway,** MD DSc FRCS FRCP FACS FACP;
former President, Royal College of Physicians and Surgeons
of Glasgow; Honorary Professor of Surgery, College of
Medical, Veterinary & Life Sciences, University of Glasgow

Every important realm of science is worthy of continuing reevaluation. The idea that a field of inquiry is "settled science" and therefore must be excluded from scientific challenge is detrimental to science. In this spirit, I am happy to recommend Michael Denton's *The Miracle of Man*. While many science books on origins focus on the question of biological evolution, others on the first cell, and others still on fine tuning in physics and the birth of the universe, Denton's latest is refreshing in the attention it pays to the astonishing degree of fitness for advanced life manifest in chemistry. Forty-five years ago, my dear friend and Berkeley colleague, the late Phil Johnson (then on sabbatical in London), quizzed me on Denton's 1985 book. I enjoyed it and encouraged him to address his powerful intellect to analyze Denton's book. Phil's study of the Denton book was perhaps the first step in the development of the intelligent design movement.

—**Henry F. Schaefer** III, PhD, Graham Perdue
Professor of Chemistry and Director of the Center for
Computational Chemistry at the University of Georgia;
Fellow of the American Academy of Arts and Sciences

"Man is the product of causes which had no prevision of the end they were achieving." Thus wrote the philosopher Bertrand Russell in perhaps the most spectacularly wrong-headed pronouncement of the twentieth century. Au contraire, in *The Miracle of Man*, Michael Denton gathers the voluminous evidence of modern science that shows the exact opposite: the universe precisely embodies the end for which it was built.

—**Michael Behe**, PhD, Professor of Biological Sciences,
Lehigh University; author of *Darwin's Black Box*,
The Edge of Evolution, and *Darwin Devolves*

If Lawrence Henderson's 1913 classic *The Fitness of the Environment* was volume 1, then Denton's 1998 *Nature's Destiny* should be considered volume 2. If one thinks that Denton completed the series with that work, one would be mistaken. In my opinion *The Miracle of Man* earns a well-deserved status as volume 3. Denton provides significant new examples of nature's prior fitness for mankind to support his anthropocentric thesis.

—**Guillermo Gonzalez**, PhD, astronomer, astrobiologist,
and co-author of *The Privileged Planet*

In his new book Michael Denton contributes a highly original new approach to the teleological design argument. Previous approaches either focused on evidence for design in the unlikely conditions of the physical constants and laws of the universe or on unlikely complex phenomena in biology. Here Denton shows that the intricate proper-ties of light, carbon, water, air, fire, and metals all are contributing to a unique prior environmental fitness of nature for human biology, which suggests that the universe is not just fine tuned for any life but was specifically designed for us and our cultural and technological development. Indeed, Denton provides powerful scientific evidence for theism and anthropocentric human exceptionalism that are at the core of the Judeo-Christian worldview. We are not insignificant ac-cidents of nature and not the cosmic orphan. Denton provides the scientific underpinning for a theistic real humanism far beyond the nihilistic implications of so-called secular humanism. This book de-serves to become a game changer that will spark a new enlightenment and re-enchantment of the cosmos in the twenty-first century.

—**Günter Bechly**, PhD in paleontology, Eberhard-Karls University of Tübingen; Scientific Curator from 1999-2016 at the State Museum of Natural History in Stuttgart, Germany; Senior Fellow of the Center for Science and Culture

The Miracle of Man is a masterful summation of Michael Denton's work. For the last several years he has been making the case that na-ture is uniquely suited—prepared in advance, as it were—to permit creatures with our physical characteristics to exist. This book brings it all together: If the characteristics of light, the sun, water, oxygen, and carbon were not precisely as they are, we would not exist. If our atmo-sphere were different, or our planet's location and composition, life as we know it could not exist. This may seem a trivial argument, but it is not. Rather than say that "hey, we are suited for the environment because we are adapted to it," we must also admit that, seen from the other side, the precise chemical properties of the elements, formed in the fires of creation, were pre-adapted to permit creatures like us—air-breathing, high energy, bipedal and terrestrial, with our basic physiology. The very chemistry of life seems to have been designed for us! Read this book if this hypothesis sounds absurd or trivial—the weight of the evidence may change your mind.

—**Ann Gauger**, PhD, biologist, Senior Fellow at the Center for Science and Culture, and co-author of *Science and Human Origins*

Michael Denton's *The Miracle of Man* is a tour de force and should feature prominently in future debates on whether humans are here on Earth by accident, as Darwinian materialism holds. This book makes clear the question is now settled, and the answer is no. We are no accident. Here is a medical doctor and biochemistry PhD with a breadth of knowledge stretching from physiology and chemistry to physics and anatomy. He cites an incredible number of well-established facts that show that Earth along with physics and chemistry were a perfect fit for human beings. Numerous facets are finely tuned and exacting, like an exquisite, fine glove that fits perfectly, a glove with a million plus fingers of all shapes, sizes, and textures. This is beyond coincidental. I like the Yogi Berra quote here: "That's too coincidental to be a coincidence."

—**Geoffrey Simmons**, MD, former Governor of the American
Academy of Disaster Medicine (AADM), and author of
What Darwin Didn't Know and *Billions of Missing Links*

In this amazing book, Dr. Michael Denton shows scientifically that nature is remarkably fit for life. Not just for life in general, but for us in particular. He describes many "ensembles of fitness" that work together to support human life and technology. The result is awe-inspiring.

—**Jonathan Wells**, PhD, biologist, Senior Fellow at the
Center for Science and Culture, and author of *Icons of
Evolution*, *The Myth of Junk DNA*, and *Zombie Science*

Michael Denton's singular achievement is to integrate the vast corpus of scientific knowledge of human physiology into a readily comprehensible coherent narrative. Along the way he reveals our existence to be a "natural miracle," governed by nature's immutable laws of the physics and chemistry of the elements of which we are made but whose combination together—and fine tuning for their purpose—lie so far beyond the realms of chance they might be termed miraculous. A powerful testimony to there being more than we can know.

—**James Le Fanu**, FRCP, physician and author of *Why
Us?: How Science Rediscovered the Mystery of Ourselves*

Some years ago, Michael Denton and I stood in the surf of one of California's spectacular beaches. I asked him, "Do you suppose the salinity levels of the earth's oceans are calibrated to foster an environment specially fit for human habitation?" "Let me look into it," he said. A few weeks later, he wrote in reply that there is indeed evidence that oceanic salinity matters to conditions for human existence and flourishing.

After reading this survey of natural history, which prepares the way for natural theology, you'll understand why I trust Michael Denton for answers to this sort of question. And you'll wonder why today's scientists are so slow to acknowledge that our terrestrial environment is exquisitely fine tuned. (Denton has wise things to say about that, as well.)

Michael Denton has compiled a resume of stunning elements in the total ensemble of fitness, and shown how the significance of each is even greater in the aggregate than it is in isolation. All of this stands ready to be observed, understood, and used by us, so the natural world and the built world are seamlessly integrated through a cognitive process exploited by humans. We can be grateful for Denton's wide-ranging curiosity and for his technical skill, both as a highly versatile scientist and as a remarkably accessible writer. *The Miracle of Man* is a stunning achievement and a wonderful capstone to his life's work.

—**R. Douglas Geivett**, Professor of Philosophy,
Talbot School of Theology, Biola University

In this marvelous book, Dr. Denton completes an epic journey through a stunning landscape of scientific discovery to arrive at the grand finale: nature's startling fitness for humankind. There are the fundamental particles that constitute our bodies—from carbon, hydrogen, nitrogen, and oxygen to various metal elements—each finely tuned to serve very precise biological functions. There is that supernaturally fine-tuned molecule of life, water, whose suite of unique properties allows it to sustain life's biochemistry and also drive the life-essential hydrological cycle. There is carbon dioxide, which warms our planet and serves indispensably in the respiratory and circulatory systems.

There are the crucial atmospheric gases, including O_3, for filtering sunlight to protect us; lightning that allows nitrogen to react with oxygen to form our proteins; and much more.

The book also touches on the wonders of the cell, and of blood and its masterpiece of a molecule, hemoglobin. Denton looks at fire and the exquisite fine tuning of the laws of physics. All of these serve man unusually well and, to my mind, suggest a miracle and more than a miracle—an intelligent mind who very much had us in mind when the cosmos came into being. That is, the universe was designed to conspire in our favor.

The Miracle of Man is a comprehensive and most convicting review of the relevant data, a survey urging any reasonable person to consider that we humans were foreseen, were planned from the beginning. The author does not insist on such a conclusion, and some of his readers may choose not to go there, but Denton's powerful new volume makes one thing undeniably clear: the data has delivered its verdict—nature is arrestingly fit for man. We are indeed, a most privileged species.

—**Marcos Eberlin**, PhD, member of the Brazilian Academy of Sciences, former President of the International Mass Spectrometry Foundation, founder of the Thomson Mass Spectrometry Laboratory, winner of the prestigious Thomson Medal (2016), and author of *Foresight: How the Chemistry of Life Reveals Planning and Purpose*

Denton backs his staggering claim that the universe is "uniquely fit" for us with a staggering weight of evidence. The search for strange forms of intelligent life elsewhere can stop now. If we find our equals somewhere else in the universe, they will have to be very much like us.

—**Douglas Axe**, Maxwell Professor of Molecular Biology at Biola University, and author of *Undeniable: How Biology Confirms Our Intuition That Life Is Designed*

CONTENTS

1. Introduction

There is, however, one scientific conclusion which I wish to put forward as a positive, and I trust, fruitful outcome of the present investigation. The properties of matter and the course of cosmic evolution are now seen to be intimately related to the structure of the living being and its activities; they become, therefore, far more important in biology than has been previously suspected. For the whole evolutionary process, both cosmic and organic, is one, and the biologist may now rightly regard the universe in its very essence as biocentric.

—Lawrence Henderson, *The Fitness of the Environment* (1913)[1]

THE YEAR 1543 WAS ONE OF THE MOST FATEFUL IN THE HISTORY of Western civilization. It was not a great battle or some dramatic shift in the balance of power on the continent of Europe. No great natural disaster was noted. No new king was crowned. But two events occurred that year, the seismic consequences of which were unimaginable at the time.

One of these set in motion an intellectual revolution that would displace man from the central place in the order of the cosmos, a place he had occupied for the previous two thousand years. The discovery would transform Western thought.[2]

In the spring of that year, the astronomer Nicolaus Copernicus lay ill in his native Poland waiting to receive from his publisher in Nuremberg the first copy of his work *De Revolutionibus*. The work contradicted

Figure 1.1. Medieval clock face featuring the sun, moon, and constellations revolving about the Earth.

long-established conventional wisdom by identifying the sun, and not the Earth, as the center of the solar system, delivering what many came to see as a profound challenge to the established worldview of the Christian West, one that since classical times had been both geocentric and anthropocentric.[3] It was a worldview immortalized in Dante's *Divine Comedy* and self-confidently displayed on the grand astronomical clocks that adorned the facades of public buildings and cathedrals in the medieval period.[4]

Also in the spring of 1543, Andreas Vesalius, a young anatomist from the leading Italian medical school of the day, was overseeing publication of his seven-volume anatomical masterpiece, *De Humani Corporis Fabrica*, among the most influential works on human anatomy ever published, and marking a quantum leap forward in efforts to accurately depict the various organ systems of the human body, work based for the first time on dissection of human cadavers rather than inference from the anatomy of animals. *De Fabrica* along with a handful of similar anatomical texts of the period revolutionized our knowledge of the human body and laid the foundations of modern scientific biology.

Neither Vesalius's work, nor Copernicus's, it should be noted, reveals the slightest connection between man and cosmos, between biology and cosmology, or any hint that man might occupy a special and preordained place in the cosmos, as was universally assumed before 1543.[5] Russian medievalist historian Aron Gurevich beautifully describes the older view that was being displaced:

> The effort to grasp the world as a single unified whole runs through all the medieval summae, the encyclopedias and the etymologies... The philosophers of the twelfth century speak of the necessity of studying nature: for in the cognition of nature in all her depths, man finds himself.... Underlying these arguments and images is a confident belief in the unity and beauty of the world, and also the conviction that the central place in the world which God has created belongs to man... The unity of man with the universe is revealed in the harmony interpenetrating them. Both man and the world are governed by the cosmic music which expresses the harmony of the whole with its parts and which permeates all, from the heavenly spheres to man. *Musica humana* is in perfect concord with *musica mundana* [the music of the universe].[6]

Indeed, so intensely anthropocentric was their vision of the world order, Gurevich explains, that "each part of the human body corresponded to a part of the universe: the head to the skies, the breast to the air, the stomach to the sea, the feet to the earth; the bones corresponded to the rocks, the veins to the branches of trees."

After 1543 that vision of reality began giving way to another. It isn't just that Copernicus's heliocentric model displaced Earth from the center of the cosmos. If one sets his work beside Vesalius's and peruses both, little if any hint of a unity between man and cosmos can be found, either by studying each work separately, or by comparing them. So, for

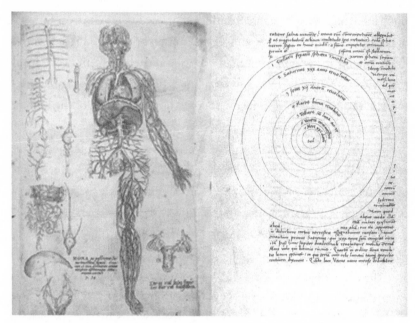

Figure 1.2. A depiction of blood vessels in the human body, from *De Fabrica* by Andreas Vesalius, and a heliocentric model of our solar system, from *De Revolutionibus* by Nicolaus Copernicus, both published in 1543.

instance, the chaotic tangle of blood vessels depicted by Vesalius stands in striking contrast to the circular perfection of the orbits of the planets as depicted by Copernicus. (See Figure 1.2.)

It was this shocking disconnect between man and cosmos—shocking in the context of a civilization that had assumed for centuries that the universe was specially ordered for human existence—which posed the basic challenge to the established conception of the cosmos as a deeply interconnected whole, with life and humankind as its end and purpose.

The effect, according to Stephen Dick, was to plunge "European thought into a crisis from which it arguably has not yet emerged."[7] The psychological shock was immense. As Alexandre Koyré comments in his classic *From the Closed World to the Infinite Universe*, man "lost his place in the world, or, more correctly perhaps, lost the very world in which he was living and about which he was thinking, and had to transform and replace not only his fundamental concepts and attributes, but even the very framework of his thought."[8] As the poet John Donne lamented, the new philosophy "Puts all in doubt... Tis all in pieces, all coherence gone."[9]

After 1543

JUST OVER half a century after *De Revolutionibus*, the Italian philosopher and mystic Giordano Bruno envisaged an infinite universe, replete with planetary homes like Earth peopled by alien beings, stretching endlessly in time and space. For this heresy (and for certain occult beliefs) he was condemned by the Inquisition, and on February 17, 1600, in Rome's Campo de' Fiori, he was burned at the stake. But the martyrdom of Bruno could not save the human-centered medieval cosmos. Already at the beginning of the seventeenth century, the first intimations of the current secular zeitgeist were already emerging, and Western man was already launched on the long journey to the nihilism of the twenty-first century.

In 1610, ten years after Bruno's death, Galileo Galilei designed and built a telescope and then turned it toward the night sky. He saw for the first time that the other planets of our solar system were indeed worlds like the Earth, and that one of these constituted an analogue in miniature of the solar system—Jupiter as a sort of mini sun and its group of moons orbiting it like so many planets. In observing the stars he also noted that many more could be seen through his telescope than with the naked eye. This led to further radical possibilities, that the stars were suns like our own but immensely far away and perhaps infinite in number. These observations, reported in 1610 in *Sidereus Nuncius*, another landmark publication in astronomy, lent Bruno's hypothetical possibility of many worlds empirical traction.

Now Earth was not only *not* the center of things, it was but one of a myriad of similar bodies possibly infinite in number. And man was perhaps just one type of intelligent being among a potential infinity of alternative forms. By the opening of the seventeenth century, the depictions of a human-centered cosmos on the medieval clocks already seemed hopelessly dated. Over the next century the growing divorce was reinforced by further advances in astronomy and biology, which provided not the slightest hint of any connection or correspondence between the two fields—between physics and biology, between man and cosmos.

The circulation of the blood described in William Harvey's *De Motu Cordis*,[10] and the tiny organisms observed by the seventeenth-century microscopists, illustrated in Robert Hooke's *Micrographia*,[11] did nothing to heal the rift. The microscope and telescope revealed two far-apart domains and provided no hint of the medieval vision of a cosmos where all things were purposefully interconnected.

The trend away from the medieval vision of the cosmos was hardly short-lived. None of the scientific advances in physics and biology right up to the mid-nineteenth century appeared to provide the slightest hint of the coherent, self-confident, human-centered world of the thirteenth century, characterized by what has been termed a "teleological and organismic pattern of thought."[12] There was nothing in Newton's laws of motion, or the series of discoveries concerning electromagnetism, or in the chemical discoveries of Priestly and Lavoisier that lent any obvious support to the old man-centered worldview. Nor did the early nineteenth-century discoveries in paleontology, comparative anatomy, biogeography, and cell theory provide any apparent reason to suppose that nature was prefigured for our existence.

After Darwin

THE FINAL act in the unraveling of the medieval human-centered worldview came in 1859 with the publication of Charles Darwin's *On the Origin of Species*. There the English naturalist argued that the only causal factor responsible for the origin of our planet's many species was the blind mechanism of natural selection working on chance variations. And if anyone missed the significance for humankind's place in his scheme, he made it explicit twelve years later in *The Descent of Man*. For many the lesson was clear: we might be the cleverest of the species, but we too were the product of blind forces that did not have us in mind, for the simple reason that, being mindless, they had nothing at all in mind.

With the acceptance of Darwinism by the biological mainstream, Western civilization took the final step back to the atomism, material-ism, and many-worlds doctrine of Democritus and other pre-Socratic philosophers of ancient Greece. As the Darwinian paradigm tightened its grip on mainstream biology and science, all vestiges of the old pre-Copernican teleological-organismic universe, all notions which placed humankind or life on Earth in any special or privileged place in the order of things, were banished from mainstream academic debate.

The implications of the final Darwinian unraveling for mainstream evolutionary biologists was memorably captured by French biochemist Jacques Monod in his materialist manifesto *Chance and Necessity*. "The thesis I shall present in this book is that the biosphere does not contain a predictable class of objects or of events," he wrote, "but constitutes a particular occurrence, compatible indeed with first principles, but not deducible from those principles and therefore essentially unpredict-able... unpredictable for the same reason—neither more nor less—that the particular configuration of atoms constituting this pebble I have in my hand is unpredictable."[13]

According to Monod the human race was adrift in an uncaring cos-mos which knew nothing of its becoming or fate, an infinite universe said to manifest not the slightest evidence of anthropocentric bias. Instead, as Harvard paleontologist Stephen Jay Gould put it, we are merely "the embodiment of contingency,"[14] our species but "a tiny twig on an im-probable branch of a contingent limb on a fortunate tree... we are a de-tail, not a purpose... in a vast universe, a wildly improbable evolutionary event."[15] Or as astronomer Carl Sagan framed the matter, "one voice in the cosmic fugue."[16]

Thus was humanity demoted to a mere epiphenomenon, to one un-purposed by-product among many, from the Imago Dei as understood in the medieval vision of humanity—that of a being made in the image of God and pre-ordained from the beginning—to a meaningless contingency, something less than a cosmic afterthought.

This modern secular vision of nature is as far removed from the anthropocentric cosmos of the medieval scholastic philosophers as could be imagined, representing one of the most dramatic intellectual transformations in the history of human thought.

But then the evidential landscape again began to change.

A Reconnection and a Second Revolution

EVEN AS the scientific vision of humankind as an accidental by-product of the cosmos consolidated its position of ascendancy in Western thought, the first seeds of a new scientific anthropocentricism were sprouting, in the Bridgewater Treatises of the 1830s. The multivolume work included such contributions as William Whewell's discussion of the striking fitness of water for life[17] and William Prout's discussion of the special properties of the carbon atom for life,[18] revealed by the development of organic chemistry in the first quarter of the nineteenth century. And ironically it was during the decades following the publication of *The Origin of Species* (1859), during the very period when Friedrich Nietzsche proclaimed that "nihilism stands at the door,"[19] when fresh scientific evidence began to accumulate suggesting that life on Earth might after all be a special phenomenon "built into" the natural order and very far from the accident of deep time and chance that the Darwinian materialist zeitgeist assumed.

These discoveries, and particularly the unique chemistry of carbon, were explored in *The World of Life* by none less than the co-discoverer with Charles Darwin of evolution by natural selection, Alfred Russel Wallace. In that 1911 work, Wallace showed that the natural environment gave various compelling indications of having been pre-arranged for carbon-based life as it occurs on Earth.[20]

Two years later, in 1913, Lawrence Henderson published his classic *The Fitness of the Environment*, which presented basically the same argument but in much more scholarly detail. Henderson not only argued that the natural environment was peculiarly fit for carbon-based life but also in certain intriguing ways for beings of our physiological design. He refers to two of the thermal properties of water, its specific heat[21] and the cooling effect of evaporation,[22] as well as the gaseous nature of CO_2[23] as special elements of environmental fitness in nature for beings of our biological design.

Building on the evidence alluded to by Wallace and Henderson, other more recent scholars, including George Wald[24] and Harold Morowitz,[25] have further defended the fitness paradigm during the twentieth century. Wald argued for the unique environmental fitness of nature for carbon chemistry[26] and photosynthesis.[27] Morowitz argued for the unique fitness of water for cellular energetics.[28]

These discoveries signal a sea change that may prove as momentous as the year 1543 one described above. This monograph will provide what is to my knowledge the most comprehensive review in print of nature's unique fitness for human biology by describing a stunning set of ensembles of prior environmental fitness, many clearly written into the laws of nature from the moment of creation, enabling the actualization of key defining attributes of our biology. The evidence puts to bed the notions of Gould, Monod, and Sagan that humankind is a mere contingent outcome of blind, purposeless, natural processes.

I agree that to claim that the findings of modern science support a contemporary take on the traditional anthropocentric worldview is highly controversial and will seem outrageous to many commentators and critics. Here a distinction may prove useful. While my *conclusions* are controversial, the *evidences* upon which they are based are not in the least controversial. In virtually every case they are so firmly established in the relevant scientific disciplines as to now be considered wholly uncontroversial conventional wisdom. In other words, the extraordinary ensembles of natural environmental fitness described in these pages, ensembles vital for our existence and upon which my defense of the anthropocentric conception of nature is based, are thoroughly documented scientific facts. What is unique to these pages is the comprehensive integration of so many disparate, if overlapping, ensembles of fitness. And when we step back from these individual groves and take in the proverbial forest in all its grandeur, the panorama, I would go so far as to say, is overwhelming.

In *The Miracle of the Cell* I showed that the properties of many of the atoms of the periodic table (about twenty) manifest a unique prior fitness to serve highly specific and vital biochemical roles in the familiar carbon-based cell, the basic unit of all life on Earth. And as I stressed, it was the prior fitness of these atoms for specific biochemical functions which enabled the actualization of the first carbon-based cell irrespective of whatever cause or causes were responsible for its initial assembly. Here the focus turns to beings of our physiological and anatomical design and the numerous ensembles of environmental prior fitness necessary for our existence. This is a prior fitness that existed long before our species first appeared on planet Earth, a fitness that led the distinguished astrophysicist Freeman Dyson to famously confess, "I do not feel like an alien in this universe. The more I examine the universe and study the details of its architecture, the more evidence I find that the universe in some sense must have known that we were coming."[29]

And it is not only our biological design which was mysteriously foreseen in the fabric of nature. As several of the following chapters also show, nature was also strikingly prearranged, as it were, for our unique technological journey from fire making, to metallurgy, to the advanced technology of our current civilization. Long before man made the first fire, long before the first metal was smelted from its ore, nature was already prepared and fit for our technological journey from the Stone Age to the present.

Prior Fitness versus Adaptive Complexity

ALTHOUGH THE primary focus of this book is to defend an updated version of the traditional anthropocentric claim by reviewing the fine tuning of the environment for our biological design, there is of course evidence throughout the text supportive of the design inference. Note, however, that the design suggested here is distinct from that highlighted in those intelligent design (ID) books and articles one usually finds at the center of evolution/design discussions. Many readers will be aware that when arguments over ID appear in the popular media they generally concern the remarkable adaptive complexity of biological systems, such as the bacterial flagellum, the mammalian eye, or the sea-going animals of the Cambrian explosion. In *The Miracle of the Cell* and here, the focus is instead on the environmental fitness in nature for cellular life and creatures like ourselves respectively, ensembles of prior fitness that stand as necessary preconditions for our existence, regardless of how one supposes the first life, animal life, or human life originated in the history of life. The case here, in other words, is based on a prior environmental fitness that made key biological adaptations possible rather than on the sophistication of those adaptations themselves.

The Human Heart

CONSIDER, FOR example, the human heart and its accompanying circulatory system. The human heart is vastly superior to any human artifact. Every second it undergoes a cycle of contraction and expansion, and beats continually and faithfully for the duration of a human lifetime. It starts beating in the womb and in eighty years will beat about two billion times. The cardiac muscle itself consists of an interconnected syncytium of billions of muscle cells specially adapted to resist fatigue and contract autonomously without external activation or control. Within the cardiac muscle cells there are trillions of tightly packed molecular arrays of contractile filaments whose regular rhythmic lengthening and shortening generate the cardiac cycle.

At rest each of us needs about a fourth a liter of oxygen per minute to satisfy our energy needs.[30] This involves the movement every minute of one hundred trillion oxygen molecules across every square millimeter of the alveolar surface of the lungs. And with every contraction the heart pumps one hundred billion red blood cells through hundreds of kilometers of tiny capillaries.[31] Coursing through the capillaries in the lungs, each of these tiny nano-machines carries one billion molecules of oxygen (O_2) from the lungs to the tissues, each loosely bound to an iron atom in the hemoglobin. By the heart's unceasing activity it ensures a bountiful supply of oxygen to provide us with the vital energy of life.

The red cells themselves, no less than the heart, are also miracles of bioengineering. During its 120-day lifetime in the circulatory system, each red cell makes hundreds of thousands of circuits, covering hundreds of miles. It is only because the red cell membranes are uniquely soft and strong—one hundred times softer than a latex membrane of comparable thickness but stronger than steel[32]—that they can withstand these repeated deformations as they squeeze though the smallest capillaries, which in many cases have a diameter of five microns, almost half the diameter of the average red blood cell.

Complex Adaptive Wonders

OUR CURRENT understanding of the heart and cardiovascular and respiratory systems, including the elegant adaptations involved in the delivery of oxygen to the tissues, did not come easily but only after more than four centuries of heroic scientific endeavor. Even as recently as the fifteenth century, erroneous views of thoracic anatomy were widespread, as witnessed by Leonardo Da Vinci's depiction of imaginary connections between major blood vessels and the airways of the lungs.[33] As we saw above, it was only in the mid-sixteenth century that Vesalius and other anatomists of Renaissance Italy provided the first relatively accurate description of the anatomy of the human lungs, heart, and blood vessels from firsthand observation and dissection, and depicted them in wonderfully illustrated anatomical texts.

But despite having for the first time an accurate picture of the basic features of thoracic anatomy, none of the sixteenth-century anatomists could have imagined what every medical student now takes for granted—the complex adaptive wonders that the cardiovascular and respiratory systems represent. There on the dissection tables in the medical schools of Renaissance Italy all was a mystery. No one grasped the physiological functions performed by the chaotic tangle of arteries and veins connecting the heart with the lungs and by the soft compressible sponge-like tissue of the lungs. Why there were blood vessels, why they conformed to the pattern they did, what arteries did, what veins did, why the arteries had muscular walls while the veins were thin and flaccid, and how the vital substance in the air (oxygen) entered the blood: all was a riddle. In Vesalius's day even the capillaries, which carry the blood from the arteries to the veins, had not been observed. And the idea that the blood carried a component of the air from the lungs to the tissues linked to an iron atom encased in a tiny particle of living matter would have seemed to them mere science fiction.

The full physiological wonder was only revealed after a long subsequent chain of discoveries. The next two great leaps forward were William Harvey's discovery of the circulation of the blood in 1628,[34] and Marcello Malphigi's discovery of the capillaries in 1661[35]; and progress continued in the twentieth century with the elucidation of the molecular structure of hemoglobin, the carrier of oxygen in the red cell, a discovery made by Max Perutz and his research group at the University of Cambridge, revealing the wondrously sophisticated phenomenon of allostery and the subtle role of the iron atom in oxygen transport.[36]

Most recently, over the past four decades, biophysical studies have revealed the robust mechanical properties of the red cell membrane, which allows the cell to survive the mechanical battering involved in the innumerable passes it makes through the microcirculation.[37] We also now understand in astonishing detail just how the oxygen released in the tissues is reduced to water by electrons flowing down electron transport chains in the inner mitochondria membrane and the counterintuitive mechanism—the chemiosmotic theory—first proposed by Peter Michell. As he explained, a stream of protons flow back across the membrane, providing the energy for the synthesis of ATP, the energy carrier of the cell.[38]

The elucidation of how the cardiovascular and respiratory systems function is one of the greatest of scientific achievements, and no one doubts that the organs, cells, and molecules—the heart and lungs, the red blood cells (hemoglobin), the electron transport chains, and many other components of these systems—are paragons of bio-engineering, fine tuned to satisfy the need to derive metabolic energy from oxidations in complex organism like ourselves.

But we now know something else, something equally if not even more remarkable, something which is the subject of this book: the adaptive wonder of the circulatory and respiratory systems and their ability to deliver 250 ml of oxygen a minute to our tissues to fuel our metabolism depends ultimately on a vast suite of diverse elements of prior environmental fitness in nature, without which all the wondrous adaptive fine tuning of the heart and circulation would be of no avail. To list and fully describe all the instances of prior fitness would fill many volumes, but to briefly mention just three here: Without the prior environmental fitness of the radiation emitted by the sun and without the transparency of the atmosphere to visual light, there would be no photosynthesis and hence no oxygen, and no oxidations in the body to provide higher organisms like ourselves with the copious amounts of energy we need to satisfy our metabolic needs. Without the prior fitness of water (with its astonishing array of unique properties) to serve as the medium of circulation, there would be no circulatory system. And without the unique prior fitness of the transition metal atoms, there would be no way to convert oxidations to metabolic energy. We will examine in depth each of these factors later in the book.

From the Start

IF THE remarkable prior environmental fitness of nature for advanced energy-hungry creatures such as ourselves were restricted to our use of oxygen for generating energy, this prior fitness would still stand as a great wonder. But as we shall see, nature also exhibits highly specific prior environmental fitness essential for many other adaptations crucial to the existence of large terrestrial organisms like ourselves, including high acuity vision, fast nerve conduction, and muscles strong enough to grant mobility to beings of our size and design.

As we will further see, nature is also uniquely fine tuned to enable beings of our biological design and size to make fire, develop metallurgy, build an advanced technological civilization, and come to do science and understand the nature of the universe and its unique fitness for our type of life.

In sum, it is as if, in an act of extraordinary prescience, there was built into nature from the beginning a suite of properties finely calibrated for beings of our physiological and anatomical design and for our ability to follow the path of technological enlightenment from the Stone Age to the present.

2. PRIOR FITNESS:
THE HYDROLOGICAL CYCLE

> For is not the whole substance of all vegetables mere
> modified water? and consequently of all animals too; all
> of which either feed upon Vegetables or prey upon one
> another? Is not an immense quantity of it continually
> exhaled by the Sun, to fill the atmosphere with Vapors
> and Clouds, and feed the Plants of the Earth with the
> balm of Dews... It seems incredible at first hearing, that
> all the Blood in our Bodies should circulate in a trice,
> in a very few minutes: but I believe it would be more
> surprising, if we knew the short and swift periods of the
> great Circulation of Water that vital Blood of the Earth
> which composeth and nourisheth all things.
>
> —Richard Bentley, *Confutation of Atheism*
> *from the Frame of the World* (1693)[1]

OF ALL THE GREAT ADVANCES DURING THE FOUR-BILLION-YEAR history of life on planet Earth, among the most consequential was the vertebrate colonization of the land followed by the grand explosion of terrestrial vertebrate diversity over the subsequent 350 million years—mice, dinosaurs, pterosaurs, albatrosses, lions, emus, pythons, killer whales, gorillas and, of course, a curious and curiously inventive biped and his many inventions.

Being terrestrial matters, most especially for *Homo sapiens*. Our scientific and technological achievements over the past several centuries depend on our being terrestrial rather than aquatic beings. It is only on dry land that fire is possible, and only by using fire can metals be extracted from their ores. Further, it is only on the land that chemical phenomena can be investigated, and only on land that electricity can be exploited for technological purposes. Thus it is that only on the land can the path to science and an advanced technological civilization be followed. And so it must be throughout the universe. With apologies to science fiction, no intelligent creature with a sophisticated technology anywhere in the universe will breathe through gills.

Moreover, none of the technological benefits of terrestrial life, nor indeed terrestrial life itself, would have been possible without a remarkable mechanism, the hydrological cycle, which delivers water and the essential minerals of life to the land. Without the continual delivery of the vital matrix of life to the land, the great continental land masses would be dehydrated and sterile, devoid of terrestrial life, and none of the advances which enabled humankind to set out on the fateful journey of technological and scientific discovery would have been possible.

The cycle itself, as it turns out, is one of the most remarkable of all natural phenomena.

Three Material States

THE HYDROLOGICAL cycle is one of the great wonders of nature, a wonder so familiar that we take it entirely for granted. Without this endlessly turning water cycle, there would be no vast herds of game on the African savannah, no great rain forests to grace the tropics, no forests of oak and ash in the temperate northern latitudes, and no possibility of humans or any other advanced terrestrial carbon-based organisms.

The hydrological cycle is enabled by several unique properties of water, the most striking of which is the existence of water (H_2O) as a solid, liquid, and gas in the ambient temperature range that exists on the surface of the Earth. No other substances on the Earth's surface, including the gases in the atmosphere and the various mineral constituents of the rocks, exist in the three forms of matter in these ambient conditions. As

Figure 2.1. Water in three forms—liquid (water), solid (the iceberg), and gas (water vapor). Clouds are clusters of water droplets condensed from water vapor.

Philip Ball comments, at Earth's surface, "nearly all of the non-aqueous fabric of our planet remains in the same physical state. The oxygen and nitrogen of the air do not condense; the rock, sands, and soils do not melt... or evaporate."[2] Within the temperature range and atmospheric pressure on our planet's surface, H_2O alone among the myriad of inorganic substances exists as a solid (ice), a liquid (water), and a gas (water vapor).[3] And it is this unique capacity of water to exist in the three multiple states of matter under ambient conditions on Earth's surface that makes possible the hydrological cycle.

Everyone learns about the hydrological cycle in school, how water evaporates from the ocean, rises into the atmosphere, cools, condenses into tiny droplets that form clouds, and coalesces into larger droplets that fall to the ground as rain or snow. Water from rain and melting snow runs into rivers and is carried back to the sea.

The sheer volume of water recycled is astounding. Every day 875 cubic kilometers evaporate from the oceans, and over the course of about

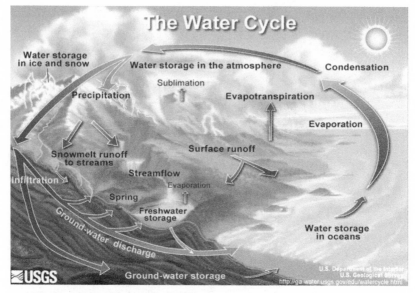

Figure 2.2. The hydrological cycle.

three millennia, a volume equal to all the world's oceans makes its way through the atmosphere and back to the sea via evaporation and precipitation.[4]

The most immediate and obvious element of fitness of this cycle for terrestrial life is that it provides a continual supply of water for land-based life forms around the globe. Only in a few particularly arid regions, such as the Atacama Desert and parts of the central Sahara, is rainfall extremely infrequent. But without the hydrological cycle, the Earth's entire land surface would be a dehydrated waste, drier and more lifeless than the driest deserts currently on Earth.

In addition to its capacity to exist in the three states of matter in ambient conditions, another property of water essential to the hydrological cycle is its relatively low viscosity and consequent relatively high mobility compared with many other fluids (and of ice compared with many crystalline solids). Its high mobility ensures its return via streams and rivers (and glaciers) to the sea to complete the cycle.

We should pause here to marvel at a fascinating aspect of the hydrological cycle, one seldom highlighted: the delivery of water to the land is carried out by and depends upon the properties of water itself. Contrast this with our artifactual designs, where key commodities such as clothes or gasoline must be delivered by extraneous delivery systems such as trucks and trains. Gasoline cannot deliver itself to gas stations nor clothes to clothing stores. But water, by its own intrinsic properties, delivers itself to the land via the hydrological cycle.

Delivering the Elements

WHILE THE hydrological cycle delivers an unending supply of water to the land, making terrestrial life and ecosystems possible, it also performs another major task indispensable to terrestrial life. The tumbling waters of a million mountain streams and rivers are continually leaching minerals from the rocks and replenishing the lakes, rivers, subsurface groundwaters, and wetlands, as well as the soil, with life-essential minerals. This vital distribution of minerals occurs thanks to several additional unique properties of water functioning in concert.

The Universal Solvent

WATER HAS long been regarded as the supreme solvent, and this outstanding quality of water is crucial to its role in dissolving the rocks and in distributing the essential minerals to terrestrial organisms. "Under the action of water, aided, to be sure, in many cases by dissolved carbonic acid," writes Lawrence Henderson, "every species of rock suffers slow destruction. All substances yield *in situ* to the solvent work of water."[5] Or as Felix Franks puts it, "The almost universal solvent action of liquid water" makes "its rigorous purification extremely difficult. Nearly all known chemicals dissolve in water to a slight, but detectable extent."[6]

The weathering action of water is greatly enhanced by two other elements of fitness: carbon dioxide is soluble in water and reacts with water to produce carbonic acid. This mild acid promotes many chemical reactions with the minerals in the rocks, making them more soluble and releasing them into the terrestrial hydrosphere. Below are two such reactions:

$$H_2CO_3 + CaCO_3 \rightarrow Ca(HCO_3)_2$$
carbonic acid + calcium carbonate → calcium bicarbonate
$$Mg_2SiO_4 + 4\,CO_2 + 4\,H_2O \rightarrow 2\,Mg^{2+} + 4\,HCO_3^- + H_4SiO_4$$
olivine + carbon dioxide + water → magnesium ions and bicarbonate
ions in solution + silicic acid in solution

By Freezing and Fracturing

TWO OTHER unique properties of water which play a critical role in the erosion and weathering of the rocks are its unusually high surface tension (second only to mercury of any common liquid[7]) and its expansion on freezing.

Working in tandem these two properties fracture rock as water, drawn into fissures due its high surface tension, subsequently expands about ten percent in volume on freezing. This exerts tremendous pressure on the surrounding rocks, splitting them open and presenting a greater surface area for chemical weathering.

Here we have another instance of the utility of water existing not only as a liquid and a gas at ambient temperatures but also as a solid. It is only because water freezes under ambient conditions on Earth's surface that this rock-fracturing phenomenon occurs. And note, the expansion of water on freezing is, like its existence in the three states of matter in ambient conditions (solid, liquid, and gas), another anomalous[8] and nearly unique[9] property of water.

Viscosity

WATER'S LOW viscosity and consequent high mobility further enhances its ability to weather rocks. Its high mobility as it tumbles over cataracts and courses down river valleys on its way to the sea, in conjunction with the abrasive action of the tiny particles of rock carried within the stream, further enhances water's erosional powers.

The viscosity of substances varies over many orders of magnitude.[10] Nitrogen's viscosity is extremely low—0.0000167 Pa-s at 0°C. (Pa-s stands for Pascal-seconds, the standard unit for viscosity measurements.) Water's viscosity is 0.001 Pa-s. The viscosity of canola oil is

0.033 Pa-s. Motor oil is 0.2 Pa-s. Tar is 30,000 Pa-s[11]. Crustal rocks are between 10^{21} and 10^{24} Pa-s.16,[12] a trillion, trillion times more viscous than water.[13]

In this enormous range of values, water's viscosity must be very close to what it is or its erosional powers would be vastly diminished. If water's viscosity were greater, closer to that of many other liquids—olive oil for example—its mobility and ability to abrade the rocks would be greatly reduced.

Working in concert, the various properties of water involved in erosion and weathering are extraordinarily efficient, wearing away immense quantities of rock remarkably quickly. Dramatic evidence of water's erosional powers is provided, for example, by the fact that Niagara Falls has retreated seven miles upstream in the last 12,000 years, a rate of erosion of one foot of solid rock per year.[14]

Glaciers

THE VISCOSITY of solid water (ice) also enhances water's fitness for eroding rocks. The viscosity of ice is extremely low for a crystalline solid. The viscosity of the rocks in Earth's outer layers is between 10^{21} and 10^{24} Pa-s. The viscosity of ice in a glacier is about 10^{11} Pa-s,[15] at least ten orders of magnitude lower. If its viscosity were closer to that of typical crustal rocks, virtually all the waters of Earth would be locked up in vast immobile ice caps at the poles and in high mountain regions. Moreover this would render ice incapable of eroding rocks, since it would lack the necessary mobility.

Liquid water also speeds up glacial flow by what's known as basal sliding. In this process, glaciers slide over the terrain they rest on, lubricated by a layer of liquid water at the base. The water is released from ice, which melts under the high pressure at the base of the glacier, greatly reducing the friction of the glacier on the ground surface beneath.[16] A similar film of water formed by the pressure of an ice skate has the same effect, reducing the friction between skate and ice, allowing the skater to glide across the surface of the ice.

As glaciers flow down river valleys or across great continental plains, they further contribute to the erosion of the lithosphere. The glaciers, in sliding over bedrock, drag rocks and rock fragments across the underlying surface, which in turn acts like sandpaper on the underlying rocks, reducing them to "rock flour." This greatly increases the area available for chemical weathering.

A Wondrous Synergy

IN SUM, the weathering and erosion of the rocks is enabled by several masterfully orchestrated chemical and physical properties of water, including but not limited to those that enable the hydrological cycle. And in passing, it is surely wondrous gratuity that this key process of provisioning land-based life with essential nutrients should lead at the same time to the creation of the magnificent mountain scenery of the great alpine chains, with their U-shaped valleys, hanging valleys, terraces, and terminal moraines.

Soil

THERE IS still more to this already extraordinary teleology. While the hydrological cycle provides both water and the necessary minerals for terrestrial life, another condition must be met for land-based life to thrive. The water enriched with the vital nutrients of life must be retained in the soil to provide an accessible and long-term supply of water and nutrients for the roots of land plants.

Amazingly, the inevitable end of water's work in weathering and eroding the rocks, the same process which yields the nutrients that fertilize and enable land-based life, also results in the formation of a set of components —sands, silts, and clays—which together make up an ideal matrix for retaining water along with its cargo of dissolved mineral nutrients in the soils.

The particulate matter found in soil provides a vast surface area and a labyrinth of micropores that retain water by capillary action. (This capillary water is in contrast to gravitational water, which resides in macropores and tends to drain rapidly from the soil.) This water-retaining property of soils is crucial to plant survival. "Because plants use water continuously, but in most places it rains only occasionally, the water holding capacity of soils is essential for plant survival," explain soil scientists Nyle Brady and Raymond Weil.[17]

So water, through the hydrological cycle, not only delivers itself and the vital elements of life to the land but also makes and delivers an ideal medium for their retention and uptake by land-based plants.

Clay

CLAYS OCCUR in most soils and contribute significantly to their water-retaining and ion-retaining properties. Clays consist of one or more clay minerals with traces of metal oxides and organic matter, and by conventional definition, are made up of fine-grained mineral particles less than two micrometers in diameter. Clay minerals are hydrous silicates or aluminosilicates, which typically form over long periods through gradual chemical weathering of silicate-bearing rocks by low concentrations of carbonic acid and other solvents.

As Brady and Weil note, the external surface area of the particles in a gram of colloidal clay is a thousand times or more that of coarse sand.[18] This contributes to clay's water-bearing properties, since the greater surface area gives water more surfaces to cling to. But that's only part of the story. Clay absorbs water and ions unusually well thanks to its unique layered microstructure, like the pages of a book but with each page a layer of silicon and oxygen atoms. These atoms carry charges, which attract water and other charged atoms or molecules (ions) so that the whole structure acts as a great reservoir, holding ions in the soil and preventing their being leached out by water as it percolates through the soil. "Next to photosynthesis and respiration," write Brady and Weil, "probably no process in nature is as vital to plant life as the exchange of ions between soil particles and growing plant roots."[19] Without it, "terrestrial ecosystems would not be able to retain sufficient nutrients to support natural or introduced vegetation."[20]

Were it not for the widespread occurrence of clay minerals in soil, vegetation on land would be greatly impoverished. There likely would be far fewer large terrestrial plants, a vastly impoverished terrestrial fauna, and probably no humans or any other large terrestrial mammals.

There is a beautiful and elegant parsimony in all this. The erosional process which draws out from the rocks the minerals and essential elements for life uses this same process to create and distribute soil, an ideal matrix for retaining water and minerals. Soil provides the means by which water and the elements can be utilized by plants, and ultimately by animals, because plants serve as intermediaries, a means to concentrate the minerals eroded from the rocks and transmit them to terrestrial herbivores and to the carnivores that prey upon them.

Water is, as Ball points out, both a "superb solvent" and in "constant flux."[21] Its mobility enables it to percolate through the soils in every corner of the terrestrial hydrosphere and, through its solvation powers, to carry a mineral harvest to terrestrial ecosystems far and wide. Water not only leaches minerals from the rocks, it conveys by its own powers the vital building blocks of life through all the soils of the terrestrial hydrosphere, fertilizing the land and enabling deserts to bloom, grasslands and forests to flourish, and animal life to thrive.

Providence

THE CONCEPTION of a global hydrological cycle, with water evaporating from the seas, condensing to form clouds, falling as rain on the land, and returning to the seas via streams and rivers to complete the circuit, dates back more than two thousand years. For many centuries through the medieval and early modern periods, many authors mistakenly thought that rainfall by itself was insufficient to account for the flow of water in rivers, and it was believed there were underground channels that conveyed water upward from the sea to springs and the source of streams. But from the early seventeenth century, rainfall was increasingly acknowledged to suffice for supplying the streams and rivers with water. Chemist John Dalton conclusively demonstrated this in a celebrated paper he presented to the Manchester Literary and Philosophical Society in 1799.[22] Yi -Fu Tuan comments that in the opening remarks for that paper, "Dalton could not refrain from expressing... his admiration for the beautiful system of nature and for the provident, unceasing circulation of water."[23]

But the water cycle was a source of wonderment and a popular topic of natural theology long before Dalton.[24] In *The Wisdom of God Manifested in the Works of the Creation* (1743), John Ray viewed the distribution of clouds and rain as "a great argument of providence and Divine disposition."[25] He also saw as providential the fact that water fell to earth via gentle drops of rain rather than pouring down from above like a cataract, since the latter "would gall the ground, wash away plants by the roots, overthrow houses, and greatly incommode, if not suffocate animals." The wind, too, he noted, plays its part in this providential arrangement, dissipating "contagious vapours" and transferring "clouds from place to place, for the more commodious watering of the earth."[26]

As Tuan explains in *The Hydrologic Cycle and the Wisdom of God*, the concept of the water cycle was from 1700–1850 the "handmaiden of natural theology as much as it was a child of natural philosophy."[27] Tuan continues:

> Of the three classes of evidence—astronomical, terrestrial and biological—terrestrial evidence proved in some ways to be the most difficult to draw upon in support of the notion of a wise and provident God. Until the concept of the hydrologic cycle was introduced and elaborated, it was difficult to argue convincingly for rationality in the pattern of land and sea, in the existence of mountains, in the occurrence of floods, etc. The hydrologic cycle served as an ordering principle, and when combined with the geologic cycle, it assumed a grandeur of inclusiveness that makes some of our modern efforts to describe the earth look like a medley of disjointed facts and ideas.[28]

A Stunning Ensemble

THE WAY that water's ensemble of elements of mutual fitness work together in the hydrological cycle to enable terrestrial life to flourish is simply stunning. This ensemble includes (1) water's unique capacity to exist in the three states in the ambient temperature range, (2) the low viscosity of both water and ice, (3) water's unique set of properties ideal for weathering and eroding rocks and for extracting from them the necessary essential elements of life while, at the same time (4), generating the key constituents of soil, which have ideal properties for retaining water enriched with a vital harvest of minerals leached from the rocks for the benefit of plant life and, indirectly, all animal life on land. This ensemble stands as a monumental testimony that nature is indeed fine tuned for terrestrial life.

Design Implications

ALTHOUGH THE main aim of this book is to review the environmental fitness of nature for beings of our biology, and not to argue for design, it is hard to imagine any ensemble of fitness more indicative of design than the way the many diverse properties of water work together in the hydrological cycle to enable terrestrial life.

It is indeed astounding that several completely distinct physical and chemical properties of water—e.g., its existing in three states in Earth's ambient temperature range, its low viscosity and high mobility, its solvation powers, its expansion on freezing, and its high surface tension—should work together so beautifully in the vital task of watering the land, breaking down rocks, weathering life-essential minerals, and creating and widely distributing water-retaining soil, essential to plant life.

Moreover, heaping wonder on wonder, one of the properties which assists in the weathering and erosion of the rocks and hence in the making of soil—water's unusually high surface tension—is also the very property which holds the water in the micropores in the soil, retaining it for utilization by plants.

Teleological Hierarchy

FINALLY, AN intriguing aspect of the way the various diverse properties of water work together to serve the end of terrestrial life is that they form what might be termed a teleological hierarchy. One unique ensemble of properties is causally prior to and necessary for the use of a second ensemble of fitness, which in turn is causally prior to and necessary for the use of a third ensemble of fitness. Thus water's unique ability to exist in three states of matter (the first ensemble) enables the hydrological cycle and is causally prior to the exploitation of a second unique ensemble of water properties, one that enables the erosion and weathering of rocks and the delivery of the vital elements of life to the land and the manufacture of soil. And this second ensemble is causally prior to and necessary for the use of a third ensemble of fitness, that which enables soil to retain water and vital nutrients and thereby allow for the growth of plants.

And recall that this remarkable teleological hierarchy is of vital utility for only a small fraction of all life on Earth. The minerals in sea water are replenished from the reaction of the upwelling magma with sea water, so the cycle is of no direct benefit to life in the sea. And while subterranean microbes may benefit from the water and minerals delivered to the land, one of the most remarkable consequences of the cycle, and vital for the growth of terrestrial plants—namely, the generation of soil—is no benefit to the great biomass of rock eaters below.

Nor is the hydrological cycle the only unique ensemble of prior environmental fitness for life on the land. As we shall see in subsequent chapters, there are other equally remarkable ensembles of prior fitness for our terrestrial way of life. Rather than being an anomaly, the striking ensemble of environmental fitness manifest in the hydrological cycle is part of a meta-ensemble of prior fitness for life on land, a vast integrated web of fitness which enables the existence of complex metabolically active terrestrial warm-blooded organisms like ourselves, and which enabled our ancestors to follow the path to our present technological civilization via fire making and metallurgy.

3. Fitness for Aerobic Life

> *The Hitchhiker's Guide to the Galaxy* describes the Earth as
> an utterly insignificant blue-green planet, orbiting a small
> and unregarded yellow sun in the uncharted backwaters
> of the western spiral arm of the galaxy. In deriding our
> anthropocentric view of the Universe, Douglas Adams
> left the Earth with just one claim to fame: photosynthesis.
> Blue symbolizes the oceans of water, the raw material of
> photosynthesis; green is for chlorophyll the marvellous
> transducer that converts light energy into chemical energy
> in plants; and our little yellow sun provides all the solar
> energy we could wish for, except perhaps in England.
>
> —Nick Lane, *Oxygen*[1]

ANYONE WHO HAS CLIMBED TO HIGH ALTITUDES IN THE HIMALA-yas will have experienced firsthand the body's compelling hunger for oxygen. Every step above 5,000 meters in the Everest valley, where the oxygen levels are about half what they are at sea level, is a considerable effort. At 8,000 meters, where the oxygen level is about one-third that at sea level, the slightest physical effort is a torture.

Most people can, after some training, hold their breath for a few minutes. Famous magician and stunt man David Blaine held his breath for seventeen minutes after extensive training and a period of forced hyperventilation and breathing pure oxygen.[2] Whales and some other deep-diving marine mammals can hold their breath for up to one or two hours. But of course even whales are forced by the same primeval hunger to eventually surface and replenish their oxygen stores.

Every tissue in the body craves and lives on oxygen, or more properly, the energy generated by the oxidation of our body's biofuels—reduced carbon compounds, including sugars and fats. When starved of oxygen after the blood supply to a tissue is suddenly cut off, such as might occur in a heart attack or stroke, the fire of life rapidly fades as the body's cells can no can longer maintain their core metabolic activities.

And it isn't just mammalian life. We share our hunger for oxygen with all other complex organisms on earth, including birds and insects as well as fish, plants, and unicellular organisms such as protozoans (e.g., *Amoeba* and the ciliate *Paramecium*).

This hunger for oxygen stems directly from a chemical universal—that the oxidation of reduced carbon compounds

CH (sugar/fat) + O_2 → CO_2 + water + Energy (ATP and heat)

yields more energy than any other chemical reaction available to carbon-based life.

One author described this as the chemical reaction "which empowers the world,"[3] because one version of it empowers the metabolism of the body and another version is responsible for the heat released when wood or coal burns,

CH (wood) + O_2 → CO_2 + water + Energy (heat),

Figure 3.1. Wildfire in California.

except that in the fiery heat of the latter, the reaction is destructively vigorous, as witnessed in a forest fire, while in the body the process is tightly controlled so the energy can be released gradually and used for our metabolic needs.

Not all organisms depend on the energy generated by "the reaction which empowers the world." There are alternative means of generating energy from various exotic chemical reactions, and a vast assemblage of anaerobic unicellular organisms obtain their metabolic energy by exploiting them.[4] Additionally, all carbon-based cells on Earth (including our own) can generate metabolic energy in the absence of oxygen by exploiting a set of chemical reactions in the glycolytic pathway that convert glucose to lactic acid and then alcohol, hence the alternative name for this series of chemical reactions, *fermentation.*

But neither fermentation nor any of the alternative non-aerobic chemical reactions provide anything like the energy of oxidations. Consequently, all complex advanced animals on earth, with their high energy requirements, use oxidation to procure metabolic energy. There are no exceptions.

As George Wald pointed out more than sixty years ago, oxidation of the body's nutrients to CO_2 and H_2O provides excess energy beyond that needed for basic processes like cell division and synthesizing the basic material constituents of the cell—excess energy that allows organisms to grow in complexity and do interesting things beyond the basic need to survive.[5] By using oxygen to burn reduced carbon compounds, organisms can not only survive but also thrive. Bees can buzz, hummingbirds hum, squids and chameleons change color, crows solve problems, and humans build rockets to the stars thanks crucially to the fact that oxidations release copious amounts of metabolic energy, far more than is needed for merely sustaining the basic metabolism of the cell.

Complexity Requires Oxygen

BECAUSE OXIDATIONS provide so much more energy for carbon-based life forms than any other available chemical reaction, the authors of a paper in the journal *Astrobiology* argue that all complex carbon-based life throughout the universe will depend on oxidations for metabolic energy. They comment:

> The reduction of oxygen provides the largest free energy release per electron transfer, except for the reduction of fluorine and chlorine.[6] However, the bonding of O_2 [where two oxygen atoms join] ensures that it is sufficiently stable to accumulate in a planetary atmosphere, whereas the more weakly bonded halogen gases are far too reactive ever to achieve significant abundance. Consequently, an atmosphere rich in O_2 provides the largest feasible energy source. This universal uniqueness suggests that abundant O_2 is necessary for the high-energy demands of complex life anywhere, i.e., for actively mobile organisms of approximately 10^{-1} to 10^0 m size scale [that is, 10 centimeters to one meter] with specialized, differentiated anatomy comparable to advanced metazoans.
>
> On Earth, aerobic metabolism provides about an order of magnitude more energy for a given intake of food than anaerobic metabolism. As a result, anaerobes [organisms that derive their energy from chemical reactions other than oxidations] do not grow beyond the complexity of uniseriate filaments of cells.[7]

They estimate that the partial pressure of oxygen in the atmosphere must exceed about 10 percent of current levels (16 mm Hg) for relatively simple aerobic organisms that rely on diffusion to reach one millimeter in size; and between 10 percent and 100 percent of current levels (between about 16 and 160 mm Hg) is needed for complex organisms with "circulatory physiology" to grow more than one centimeter in size.

The correlation between oxygen levels in the atmosphere and biological complexity is further evidenced, as David Catling and his colleagues explain, by the fact that during the history of life on Earth the size and complexity of organisms during geological time correlate well with the partial pressure of oxygen (pO_2) in the atmosphere and oceans. As they note, oxygen levels rose from only traces in the primeval oceans to current levels about 500 million years ago, coincident with the advent of complex metazoan life in the Cambrian era, after which levels have remained between 15 and 30 percent (21 percent at present).

From the above one thing is clear: If there were no atoms in the periodic table with the chemical vigor of oxygen, then carbon-based life on Earth would be imprisoned forever at the level of primitive unicells. The ascent of life from a putative protocell to the emergence of humankind was in a real sense gifted by the prior fitness for energy generation of this very special atom. Without the prior fitness of oxidations for bioenergetics, our biosphere would be no more than a thin layer of slime made up of bacterial biofilms.

The chemical vigor of the oxygen atom, then, is an indispensable element of prior environmental fitness for advanced biological life.

Finally, an intriguing implication of the chemical reactivity of oxygen is that it cannot persist for long in any planetary atmosphere unless it is continually replenished by some complex biochemical process akin to photosynthesis.[8] And for this reason its presence in the atmosphere of an exoplanet is widely assumed to be a reliable biosignature.[9] Moreover given, as we have seen above, that all intelligent life anywhere in the universe will depend on the energy supplied by oxidations, its presence in the atmosphere of an exoplanet may also serve as a biosignature for intelligent life there.

Sunlight

THE OXYGEN we breathe as well as the reduced carbon compounds such as sugars and fats we oxidize in the body to produce our metabolic energy are generated by photosynthesis in the chloroplasts of green plants. The light of the sun provides the necessary energy for this process. Without it, there would be no photosynthesis. This is well understood, but what is not widely appreciated is that the fitness of the sun's radiation for life depends on several extraordinarily improbable coincidences in the nature of things.

The Electromagnetic Spectrum

The electromagnetic spectrum (EMS) is comprised of many forms of electromagnetic radiation. The various types flow through space as energy waves akin to ripples on the surface of a lake. And just as the waves on a lake may have different wavelengths—small ripples may be only a fraction of an inch from crest to crest while large waves might be a foot or more from crest to crest—so too do the wavelengths of the various kinds of electromagnetic radiation vary, though over a far, far greater range.

The two most familiar forms, and the types of radiation that our sun primarily emits, are visible light and infrared (IR) radiation, the latter commonly experienced as heat. (We experience infrared radiation as warmth when the sun shines on the skin.) But there are various other kinds of electromagnetic radiation, such as radio waves, microwaves, ultraviolet rays (UV), X-rays, and gamma rays. The reason that the effects and uses of the types of radiation in different regions of the electromagnetic spectrum are so varied is that their differing wavelengths interact with matter in very diverse ways.[10]

The Visual Band

There is only one tiny band of radiation in the inconceivably vast range of the electromagnetic spectrum with the right energy levels for photochemistry, and this coincides primarily with the visual region. The more energetic, shorter wavelength electromagnetic radiation in the gamma, X, and most of the ultraviolet regions of the spectrum strip electrons from atoms and molecules, destroying the delicate macromolecules in the cell. The less energetic, longer wavelength radiation—in the infrared, microwave, and radio wave regions—is generally too weak to activate matter for chemical reactions.[11] The region of the spectrum where the wavelengths have the just-right energy for gentle controllable photochemistry such as occurs during photosynthesis is the visual band, along with the near ultraviolet and near infrared wavelengths closely adjacent to the visible band.

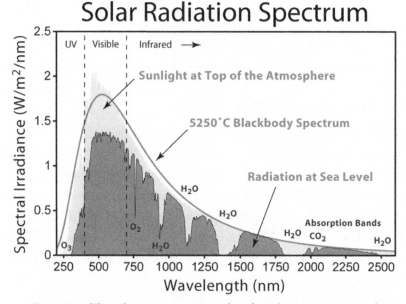

Solar Radiation Spectrum

Figure 3.2. The solar spectrum. Note that the solar spectrum extends somewhat further into the infrared region than shown. Other depictions show the spectrum extending to about 4,000 nanometers.

Now this might not seem so remarkable until one reflects on the vast range of wavelengths in the electromagnetic spectrum. As I wrote in *Children of Light*, "The total range of wavelengths in the EM spectrum is inconceivably vast. Some extremely low-frequency radio waves may be a hundred thousand kilometers from crest to crest, while some higher-energy gamma waves may be as little as 10^{-17} meters across (only a fraction of the diameter of an atomic nucleus). Even within this selected segment of the entire spectrum the wavelengths vary by an unimaginably large factor of 10^{25} or 10,000,000,000,000,000,000,000,000."[12]

To gain some sense of the size of this figure, note that the number of seconds since the beginning of the universe fourteen billion years ago is substantially less than 10^{18} seconds. To reach 10^{25} we would have to count off seconds day and night through a period equal to more than twenty million times the age of the universe.

It's in such a tiny region of the electromagnetic spectrum where the photochemical action is, chemistry essential to all advanced life on Earth, and this brings us to a marvelous instance of prior environmental fitness for creatures like ourselves: the sun emits nearly half of its radiation in this miniscule visual region of the spectrum, between wavelengths of about 380 to 750 nm in length.

The Infrared Band

The sun does something else as well. The other half of its radiant output is primarily in another equally tiny region of the electromagnetic spectrum, one immediately adjacent to the visual region—between wavelengths of 750 to somewhat beyond 2,500 nm, a swath within the larger infrared range. This radiation provides about half the vital heat to warm our planet's atmosphere. Without it Earth's entire surface would be a frozen wilderness far colder than the Antarctic. It is thanks to the heat of the sun (and to our atmospheric gases absorbing this heat) that water exists in liquid form on Earth's surface and the average global atmospheric temperature is maintained well above freezing, in a temperature range which enables the chemistry of life to proceed.

So the radiant output of the sun is fit for terrestrial life, including creatures like ourselves, in two vital ways: (1) The sun puts out the right light for photosynthesis, enabling photosynthesis and thereby giving us our foodstuffs as well as the vital oxygen we need to burn those same foodstuffs in the body. And (2) the sun puts out the right radiation to warm us, raising Earth's average temperature to about 15°C, well above freezing and enabling life to thrive.[13]

Starlight

The radiant output of a star depends on its surface temperature, which in the case of the sun is close to 6,000°C.[14] At this temperature most of the radiant emission takes the form of visual light and heat.[15] It is not just our star, the sun, that emits most of its radiation in the visual and IR. Although for stars substantially hotter than the sun, most of their stellar output is in the ultraviolet regions and higher, which is harmful to life, such stars represent only a small fraction of all stars in the cosmos. Most stars have surface temperatures below 6,000°C, and, like our sun, emit nearly all their radiation as visible and infrared light. What this means is that the entire cosmos is bathed in the "light of life."

That the radiation from the sun (and from most stars) is concentrated into a tiny band of the electromagnetic spectrum that provides precisely the two types of radiation required to maintain life on Earth— visible and infrared light—is surely a remarkable coincidence. And this is a genuine coincidence, as the compaction of solar radiation into the visible and near infrared is determined by a completely different set of physical laws from those that dictate which wavelengths are suitable for life and photosynthesis. The coincidence was described as "staggering" by Ian Campbell in *Energy and the Atmosphere*:

> It just happens that the energy required to excite an electron bound within a molecular orbital to one of the empty orbitals at higher energy is precisely of the order of 10^{-19} to 10^{-18} J [where J = joules]. The important consequence is that the solar radiation impinging upon a terrestrial molecule can be absorbed in a primary photochemical act and the electromagnetic energy becomes converted into potential chemical energy available to induce secondary reactions. Moreover the typical energy required to break chemical bonds in molecules lies in much the same range... Hence visible light, as it may be termed generally, has exactly the right scale of energy per light quantum or photon to give rise to the possibility of photochemistry, that is chemical reactions driven by the energy of sunlight.[16]

As he further notes, because the sun's surface temperature is what it is, its maximum intensity of photon output in the continuous spectrum is a wavelength of about 600 nm, and the max output in terms of energy happens at about 450 nm, in the visual range. That solar radiation is just right for photochemistry, he comments, "is the more staggering when we realize that temperatures well above one million degrees Kelvin must exist within the sun below the visible surface!... What chance then for life to have evolved had the bulk of the sun's energy output been of quanta in the energy ranges 10^{-20} to 10^{-21} J, or 10^{-18} to 10^{-17} J, or alternatively had the energies required to excite electrons in molecules or break chemical bonds been in the same ranges?"[17]

The light of the sun is fit for advanced life on a planetary surface in another way. As I explained in *Children of Light*:

> The life of stars like our sun is measured in billions of years. The sun will continue to put out radiation in the visual and IR bands—the right light for life—for another four billion years. This means that life on the surface of the Earth (and on any rocky planet in the habitable zone circulating a typical star) is ensured an ideal source of energy over enormous time spans, which, while beyond our ordinary experience, are necessary if life is to develop and thrive on the surface of a planet over periods of millions of years. Moreover, there is also very little day-to-day, year-to-year change in the solar output—another crucial element of fitness without which the Earth's climate would be unstable and unsuited for the thriving of complex, advanced life.[18]

Notice that stars about the mass of the sun endure for many billions of years (crucial for the development of advanced planetary life), and they possess the right surface temperature to emit the life-friendly radiation (in the visual and infrared region of the spectrum). As Carlos Bertulani notes, even a star just three times the sun's mass has a vastly shorter lifespan, about 370 million years. That's far less time than it took on Earth for the oxygen levels in the atmosphere to reach levels where they could support advanced aerobic life.[19] On the other hand, stars of less mass than our sun have longer lifetimes but, being cooler, put out less visual light as a proportion of their total radiance.[20]

The Atmosphere

FOR PHOTOSYNTHESIS to proceed on a planet like Earth, sunlight (visual light) must penetrate the atmosphere all the way to the ground, and part of the sun's infrared radiation needs to be absorbed by the atmosphere so as to warm the planet into the ambient temperature range, where the chemistry of life, including the chemistry of photosynthesis, can work its magic.

Happily, our atmosphere obliges. Earth's atmosphere absorbs a significant fraction of the infrared radiation—warming the atmosphere into the ambient range—and lets through nearly all of the radiation in the visual region to empower the process of photosynthesis.

Some infrared radiation does reach the Earth's surface, felt as warmth on the skin, and some penetrates a little way into water, as is commonly experienced in a swimming pool. But there are several major atmospheric absorption bands in the near infrared region that capture and retain the sun's heat, raising our planet's surface temperature by 33°C over what it would be without them, a chilly −18°C.[21]

If our atmosphere didn't absorb at least a significant fraction of the infrared radiation when the sun was shining, the atmosphere would be intolerably hot during the day, and when night fell the temperature would plunge below zero. We would experience wild temperature swings like those on the moon. There temperatures spike in the daytime to more than 100°C (the boiling point of water at sea level) and plunge to −178°C at night, a temperature far far colder than any experienced on Earth today.[22] This wide variation is because the moon has no atmosphere to retain heat at night or prevent the surface from getting so hot during the day. No type of carbon-based plant life based in a water matrix could survive such massive temperature fluctuations.

On the other hand, if our atmosphere absorbed too much in the infrared region, that too would be disastrous. And this highlights another intriguing element of fitness in the absorption pattern of electromagnetic radiation in the infrared region. The windows between the absorption peaks are as crucial as the peaks. Why? Because without some spectral windows, all the infrared radiation would be absorbed by the atmosphere, none could be radiated back out into space, and Earth would suffer a runaway greenhouse effect, ending up a hellish hothouse like Venus.

In this context an intriguing feature of our atmosphere's absorption spectrum is a sizable absorption window between eight and fourteen microns. It's intriguing because the sun is not the only body that emits infrared radiation. The Earth also does, since all bodies at a given temperature emit radiation with a characteristic range of wavelengths. In Earth's case, the emission peak is in the infrared region near 10 microns.[23] And our atmosphere's absorption gap allows a significant fraction of Earth's infrared emission to escape into space through the eight-to-fourteen micron window. Around a fourth of the outgoing infrared emission from Earth escapes through this window,[24] which consequently plays a major role in preventing our planet from going the way of Venus. If all radiation in the infrared between 0.80 and 100 microns had been absorbed by the atmospheric gases, if there were no windows, a runaway greenhouse would have been inevitable. The Earth would be a hot, Venus-like planet. Upon these windows, including the eight-to-fourteen-micron window, all advanced life on the surface of the Earth, including of course *Homo sapiens*, depends.

It is no exaggeration to say even with all the other elements of fitness that make possible our existence, without this eight-to-fourteen micron window—but one small detail in the atmosphere's overall absorption spectrum—we wouldn't exist. This represents yet another stunning instance of the biocentric fine tuning of nature.

Before turning to the role of specific atmospheric gases in the fine-tuning of our atmosphere for advanced terrestrial life, a few quick notes on some additional fortuities regarding Earth's relationship to light.

The light that passes through our atmosphere must penetrate water, not just to gift the sun's energy to aquatic plants but because water is the matrix of life, and to reach the chloroplasts in any green plant, aquatic or terrestrial light must traverse the water in the cell. Again nature obliges as water is transparent to radiation in the visual band as a liquid, as a vapor in the atmosphere, and as ice.[25] If liquid water or water vapor in the atmosphere absorbed visual light—the right light for photosynthesis— then photosynthesis would not be possible, and Earth would be devoid of aerobic life forms.[26]

Also fortuitous is the transparency of our atmosphere to visible light, which made important scientific advances possible, as Carl Sagan underscored in his 1980 book *Cosmos*. There he asked us to imagine intelligent life evolving on a cloud-covered planet such as Venus. "Would it then invent science?" he asked. "The development of science on Earth was spurred fundamentally by observations of the regularities of the stars and planets. But Venus is completely cloud-covered... nothing of the astronomical universe would be visible if you looked up into the night sky of Venus. Even the sun would be invisible in the daytime; its light would be scattered and diffused over the whole sky—just as scuba divers see only a uniform enveloping radiance beneath the sea."[27]

Finally, it is not just that our atmosphere lets through the right light. It also strongly absorbs radiation from the dangerous or potentially dangerous regions of the electromagnetic spectrum on either side of the visual and near infrared regions.

The Atmospheric Gases

ANOTHER REMARKABLE aspect of the absorption characteristics of Earth's atmosphere is that it arises from the combined absorption spectra of the atmospheric gases, five of which—nitrogen (N_2), oxygen (O_2), ozone (O_3), carbon dioxide (CO_2), and water vapor (H_2O)—are bound to be present in the atmosphere of any planet hosting complex carbon-based biological life. It is their combined absorption characteristics which lets through just the right light for photosynthesis while at the same time absorbing just the right amount of heat, as well as most of the harmful radiation outside of the visual and infrared regions.

Oxygen

As we've seen, oxygen (O_2) is indispensable to complex organisms such as ourselves. We need a lot of it (250 ml every minute, even at rest). Indeed, the metabolic rates needed to sustain the most advanced biological life depend on taking oxygen directly from an atmosphere (see the next chapter). Atmospheres sustaining complex aerobic life will inevitably contain substantial quantities of oxygen (see above).

Ozone

Where there is O_2 in an atmosphere there is bound to also be ozone (O_3), since it's formed in the stratosphere by the reaction of individual oxygen atoms with molecules of dioxygen, catalyzed by the action of UV light.

$$O_2 + O = O_3$$

Ozone is important to life because it absorbs harmful ultraviolet radiation.

Carbon Dioxide

Breathing involves taking in oxygen and exhaling carbon dioxide (CO_2), which is a major product of aerobic metabolism (the process which provides us with 90 percent of our energy needs). Consequently CO_2 will be found in the atmosphere of any planet where organisms use the oxidation of reduced carbon to generate energy. Carbon dioxide is also essential to plants, which require it for photosynthesis. Moreover, CO_2 is the only feasible carrier of the carbon atom to all parts of any carbon-based biosphere.

CO_2 is also delivered to the atmosphere on Earth by volcanic activity and is recycled via silicate weathering.

Water Vapor

Atmospheric water vapor will be found in the atmosphere of any planet harboring abundant carbon-based life because water is the essential physical matrix of all carbon-based cells and, as we'll see, the necessary medium of the circulatory system in all complex multicellular organisms. Only worlds that possess water can harbor carbon-based life, hence the NASA adage "follow the water" in searching for extraterrestrial life. And since water evaporates at temperatures fit for biochemistry, some water vapor is bound to be present in the atmosphere of any world bearing carbon-based life.

Nitrogen

Atmospheric nitrogen provides most of the nitrogen atoms incorporated into organic compounds by life on Earth. It's one of the four core atoms of organic chemistry alongside carbon, oxygen, and hydrogen. It provides necessary density to the atmosphere, keeps our oceans from evaporating, and serves as a fire retardant, slowing the speed that fire spreads, rendering it controllable. Nitrogen is the only viable candidate for these roles and thus appears to be an essential ingredient in the atmosphere of any planet hosting carbon-based life.

All this suggests that oxygen, nitrogen, water vapor, carbon dioxide, as well as ozone are bound to be present in the atmosphere of any world inhabited by oxygen-utilizing, advanced carbon-based life, for reasons over and above their life-friendly atmospheric transparency for the right kinds of electromagnetic radiation.

The Right Proportions

OUR ATMOSPHERE not only has the right components for complex aerobic life, it also has them in the right proportions. Only an oxygen concentration of about 20 percent, at a partial pressure of more than 80 mm Hg provides the requisite oxygen for the active metabolism of organisms like ourselves (discussed in the next chapter). If the concentration were substantially higher, fires would be a far greater danger. In the case of nitrogen, only a considerable quantity of nitrogen provides the density and pressure needed to keep fires from raging uncontrollably in oxygen-rich atmospheres such as Earth's, and to prevent the oceans from evaporating.

CO_2 levels have varied throughout geological time, although over the past 400 million years—since advanced life colonized the land— they have almost certainly never reached levels ten times those of today and probably never more than about four to five times present levels.[28] A recent study provided evidence of this. It found that raising CO_2 levels in controlled atmospheres up to four times present levels diminished cognitive function in human subjects.[29] This gives some indication of a CO_2 ceiling, beyond which advanced life may no longer viable.

The Greenhouse Gases

DIATOMIC MOLECULES with the same two atoms, such as O_2 or N_2, do not absorb infrared radiation. This is quite fortunate for life on Earth, since if either of these two gases, which make up most of our atmosphere, were strong absorbers of infrared radiation, Earth likely would have become a boiling cauldron like Venus, with temperatures hot enough to melt lead.[30]

Also fortuitous: the major greenhouse gases CO_2 and H_2O are both stable in the presence of O_2. This is enormously important. If they were unstable in the presence of oxygen, the whole atmospheric system and global heat balance would collapse. Aerobic life, our sort of life, would be impossible. However, in keeping with nature's profound fitness for advanced life as it exists on Earth, H_2O and CO_2 are fully oxidized and stable in the presence of oxygen. Nitrogen, the major component of the atmosphere, is also stable in the presence of oxygen, because the nitrogen atoms in N_2 bond strongly with each other and resist combining with oxygen. The stability of water, carbon dioxide, and nitrogen in the presence of oxygen is a point worth underscoring, since most other substances (apart from the noble gases) react strongly with oxygen—in some cases, explosively.

A fascinating further teleological aspect to all this concerns the quantity of ozone in the atmosphere. Because of the vast amounts of O_2 in the atmosphere, inevitably there will also be some ozone (O_3). Although ozone is indispensable for blocking harmful ultraviolet radiation, it is also a powerful greenhouse gas that absorbs strongly in the infrared region—one thousand times more strongly than CO_2.[31] Because of this, anything beyond trace amounts of ozone would contribute dangerously to the greenhouse effect. This means that its life-giving fitness in absorbing the dangerous ultraviolet radiation between 0.20 and 0.30 microns would be negated entirely if more than trace amounts were necessary for that vital task, or if it were produced in excess amounts by the action of ultraviolet radiation on O_2 in the stratosphere. Happily, only trace amounts are needed to effectively block harmful ultraviolet radiation, and the rate of breakdown of ozone in the stratosphere almost equals its rate of synthesis, guaranteeing that it is indeed only present in trace amounts.

Finally, an intriguing aspect of ozone's synthesis in the atmosphere is that ozone (O_3) and diatomic oxygen (O_2) indirectly promote their own formation by absorbing dangerous ultraviolet radiation and thereby protecting plant life, both aquatic and terrestrial, which synthesize the oxygen from which ozone is formed. This is yet another beautiful example of the parsimony and elegance of nature's stunning fitness for aerobic life.

Vital Coincidences

THE ABSORPTION properties of our atmosphere are not vital for all carbon-based life on Earth, but are for plants and energy-hungry aerobes like ourselves. Our atmosphere's fortuitous mix of gases enables photosynthesis and the manufacture of oxygen, warms Earth into the ambient temperature range, and shields life from harmful radiation. Even slight differences in our atmospheric gases' absorption properties, or in their relative concentration, and Earth would be uninhabitable, particularly for aerobic life. And note, these gases exist in our atmosphere, and in the proportions they do, because of factors quite distinct from the life-essential absorption properties described above.

There is a final twist to this teleology: Three of the key atmospheric gases whose physical absorption properties are indispensable to the process of photosynthesis are also central players in the process of photosynthesis itself.

$$6CO_2 + 6H_2O + \text{light} + \text{heat} \rightarrow C_6H_{12}O_6 + 6O_2$$

Indeed, they are the major reactants in the process. It is as if CO_2, H_2O, and O_2 were deliberately colluding to incorporate themselves into the stuff of living matter.

Light and Air

LET'S REVIEW. The laws of nature, which determine the absorption properties of the atmospheric gases, have no logically necessary connection with their chemical properties or the chemical properties of their constituent atoms, which are of such utility to life. This is a striking fortuity in the nature of things.

Similarly, there is no connection between the laws of nature which determine the tiny size of the biologically useful region in the electromagnetic spectrum, and those laws which determine the radiant output of the sun. And there is no connection between the radiant output of the sun and the laws determining the absorption properties of the atmospheric gases and liquid water.

So here we have several coincidences on which the existence of oxygen-hungry aerobic organisms like ourselves depends. In the fifteenth edition of the *Encyclopaedia Britannica*, in the article entitled "Electromagnetic Radiation," the authors comment, "Considering the importance of visible sunlight for all aspects of terrestrial life, one cannot help being awed by the dramatically narrow window in the atmospheric absorption... and in the absorption spectrum of water."[32]

And it isn't just the "dramatically narrow window." We should be in awe of the entire ensemble of prior environmental fitness summarized in this chapter, an ensemble that enables photosynthesis and, by extension, our own existence as oxygen-hungry "light eaters."

Simply put, our existence, inhabiting the surface of a planet like Earth, deriving energy generated by the oxidation of the reduced carbon compounds manufactured during the process of photosynthesis, depends on what can only be described as an extraordinarily improbable degree of environmental fitness in the order of things. Note, too, that the improbable coincidences reviewed above are largely irrelevant to the other major domain of carbon-based life on our planet—the great biomass of "rock-eating" anaerobic denizens of the dark.[33] Nature's awe-inspiring fitness for photosynthesis is a fitness for our type of life, for life in the light, for life on a planetary surface, for creatures such as ourselves.

These coincidences should inspire both awe and wonder. They provide compelling evidence of a very special fitness in nature for the generation of oxygen for oxygen-hungry beings like us.

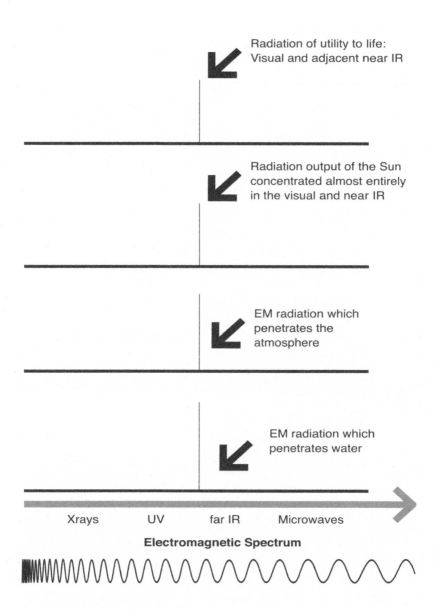

Radiation of utility to life:
Visual and adjacent near IR

Radiation output of the Sun
concentrated almost entirely
in the visual and near IR

EM radiation which
penetrates the
atmosphere

EM radiation which
penetrates water

Xrays   UV   far IR   Microwaves

Electromagnetic Spectrum

Figure 3.3. Coincidences—the narrow windows in the EM spectrum
that facilitate photosynthesis.

4. Prior Fitness: The Atmosphere

> We use about 2 milliwatts of energy per gram—or some
> 130 watts for an average person weighing 65 kg., a bit
> more than a standard 100 watt light bulb. That may not
> seem a lot, but per gram it is a factor of 10,000 more than
> the sun (only a tiny fraction of which, at any one moment,
> is undergoing nuclear fusion). Life is not much like a
> candle; more of a rocket launcher.
>
> —Nick Lane, *The Vital Question*[1]

WE SAW IN THE PREVIOUS CHAPTER THAT BECAUSE THE ENER-
gies of oxidations far surpass that of any other available reactions, oxygen is vital to all complex carbon-based life. We also reviewed
a remarkable ensemble of fitness for the vital process of photosynthesis
by which green plants use the energy of sunlight to manufacture the reduced carbon compounds and the oxygen we use to burn them in the
body to generate our metabolic energy. In this chapter we review another
ensemble of fitness which enables terrestrial aerobic life to thrive in an
atmosphere enriched with copious quantities of oxygen.

Lungs versus Gills

ALL ORGANISMS with the highest metabolic rates (rates of oxygen consumption)—mammals, birds, and flying insects—possess adaptations
to obtain their oxygen directly from the atmosphere. Mammals use a
bellows type of lung. Birds, a one-way "through put lung." Insects, a tracheal system of microtubes which permeate their tissues.[2]

Figure 4.1. Purple-throated carib hummingbird.

The rate of oxygen consumption of some air-breathing organisms is astounding. As a hummingbird darts from flower to flower, hovering as it sucks up its reduced carbon fuel in the form of sugary nectar, its wings may beat eighty times a second, creating the characteristic humming sound and hence the name of these delightful birds. To supply its wing muscles with sufficient oxygen, its heart rate may reach more than a thousand beats per minute.[3] During flight, oxygen consumption per gram of body mass is about ten times higher than in that of elite human athletes.[4]

But even the hummingbird's stunning rate of oxygen use is exceeded in the muscles of flying insects. A bee's flight muscles, for example, use oxygen at about three times the rate of a hummingbird's.[5]

No fish or any other water-breathing organism using gills or their equivalent to extract oxygen from water can come close. This is because of a fundamental constraint: it is far more difficult to obtain oxygen from water than from air, limiting the metabolic rate of aerobic water-breathing organisms. John Maina explains:

> As a respiratory medium, air is a more cost-effective respiratory fluid: water is 50 times more viscous than air; the concentration of dissolved

oxygen in water is about one-thirtieth that in air; the rate of diffusion of oxygen in water is lower by a factor of 8000 compared with that in air; and the capacitance coefficient, i.e. increment of concentration per increment in partial pressure of oxygen, in water is 30 times lower in air. In saturated water, at 20°C, 1 mL of oxygen is contained in 200 g of water while 1 mL of oxygen is present in 5 mL of air.... All other conditions being equal, owing to greater viscosity of water, compared with air breathing, water breathing requires more energy to procure an equivalent amount of oxygen.[6]

Goran Nilsson underscores the same point: "A water breather will have to move about 20,000 times more mass over its respiratory surface than an air breather to access the same amount of oxygen."[7]

Because the cost of breathing is far greater in a fish than in air-breathing organisms, the maximal metabolic rates of terrestrial aerobic organisms far exceed those of the highest recorded for fish. While the resting metabolic rate of a tuna (a fish with one of the highest metabolic rates in the sea[8]) approaches that of humans, the maximum oxygen consumption of a tuna is four times less than that of a human, sixty times less than that of a hummingbird, and nearly 200 times less than the flight muscles of a bee.[9]

The constraints imposed by water breathing on vertebrate physiology are also illustrated by the fact that despite the great advantages of being warm blooded (see Chapter 7), no water-breathing organism other than the opah, a type of moonfish, has achieved genuine whole-body endothermy (warm-bloodedness).[10]

The metabolic cost for water breathers is simply too great. Heat generated in an organism is a product of the oxidation of the body's reduced carbon fuels.

Oxygen + Carbon-Hydrogen compounds → CO_2 + H_2O + heat

To maintain the body temperature above that of an animal's surroundings requires a high rate of oxidation (a high metabolic rate), and water breathing is simply too inefficient to provide oxygen at the required rates. For aquatic organisms this problem is compounded by the fact that water is a better conductor of heat than air is. Not unsurprisingly, warm-blooded aquatic organisms, such as whales, dolphins, and porpoises, all breathe air.

Although recent research has revealed that many fish have surprising abilities,[11] and some exhibit a remarkable degree of intelligence, their far lower metabolic rates preclude their developing the far larger energy-demanding brains with the cognitive abilities of clever air-breathing animals, including dolphins, ravens, chimpanzees and, of course, humans. And even on some alien world in which gilled aquatic creatures somehow had managed to reach metabolic rates and intelligence equal to those of their "terrestrial brothers," they could *still* not develop a technology. Using fire, developing metallurgy, having knowledge of chemistry, and using electricity for technology (water is a conductor, air an insulator) are all impossible in water. Our hypothetical counterfactual highly intelligent fish or octopus would still be stuck in an eternal Stone Age, able perhaps to use primitive tools but unable to use fire, make metal tools, gain knowledge of chemistry, or invent electrical devices.

If you want to possess a high metabolic rate, be warm-blooded with fast neural processing, fast thinking, and fast reflexes; if you want to possess an oversized energy-hungry brain like that of modern humans, and if you want to successfully pursue technological development, you must be an air-breather. And as we will explore in Chapter 7, it seems likely that any carbon-based life form anywhere else in the universe intelligent enough to develop an advanced technological civilization also will be a warm-blooded air-breather.

Oxygen Being Gaseous

FOR US to take in oxygen from an atmosphere, oxygen must of course be a gas in ambient conditions. Indeed, if oxygen were a liquid or solid in ambient conditions, taking such a reactive atom into the body in such a concentrated form without corroding the organs responsible for its uptake would prove impossible, even allowing for a lung redesign. Such a scenario has never been seriously proposed, even allowing for some counterfactual organ adapted for such a feat.

Intriguingly, the existence of an oxygen-enriched atmosphere depends on a property of oxygen itself which is seldom acknowledged—the low solubility of oxygen in water, far lower than most other gases.[12] And this means, as David Catling and his colleagues point out, "Oxygen can be freed into an atmosphere and is sufficiently soluble (0.2 mM in 25°C seawater) to be distributed throughout an ocean. At the same time, O_2 is sufficiently insoluble that on Earth it partitions between the atmosphere and ocean in a 140:1 ratio, allowing the existence of complex life on land."[13]

By preventing oxygen from being swallowed by the oceans, oxygen's very low solubility allows it to accumulate in vast quantities in the atmosphere for the crucial benefit of terrestrial air breathers. Note, however, that since water is the matrix of life, in which all the biochemistry of the cell takes place, unless oxygen was soluble to some degree, its powers could never be used by either terrestrial or aquatic aerobic organisms. Another possible benefit of the low solubility of oxygen is that it lessens the concentration of harmful oxygen free radicals in the tissues and in bodies of water exposed to the atmosphere. This is a topic we will return to in a later chapter.

Thin Air

FOR AIR-BREATHING to enable advanced terrestrial organisms to acquire sufficient amounts of oxygen to supply their metabolic demands, there must be copious amounts of oxygen in the atmosphere. And not all oxygen-containing atmospheres provide sufficient oxygen to supply the energy needs of birds or bees or indeed our own need for 250 mls per minute of pure oxygen even at rest.

One such atmosphere is the summit of Mount Everest. There the partial pressure of oxygen (pO_2) in the atmosphere (the proportion of atmospheric pressure contributed by oxygen) is less than one-third that at sea level.[14] Although it is possible to survive for a short period of time

Figure 4.2. Mount Everest.

even on the summit of Everest (8,848 meters), the low oxygen partial pressure is insufficient for long-term survival. Starved of sufficient oxygen to generate the required energy levels to maintain normal physiological functions, the tissues begin to deteriorate and life ebbs away. As the author of a recent review comments, "While some individuals can make the 8,848 m ascent of Mount Everest without breathing supplemental O_2... man cannot live and reproduce at altitudes of ~>5,000 m for more than short periods of time... Basically, every organ is affected by such degrees of hypoxia, with loss of lean body mass, and cerebral, gastrointestinal and reproductive disturbances that progress to intolerance for that altitude."[15] This is why high altitudes in the Himalayas are properly designated by climbers as "the death zone."

Currently there is no permanent human habitation even as high as the Everest base camp at 5,400 meters. The highest is a mining town in the Peruvian Andes—La Rinconada at 5,100 meters.[16] At this altitude the pO_2 is 83 mm Hg,[17] slightly more than half the pO_2 of 160 mm Hg at sea level, and while the record for the highest permanent human habitation may be partially set by economic factors,[18] it probably represents a figure close to the maximum compatible with long-term survival.[19]

Native Tibetans and the inhabitants of the high Andes exhibit adaptations to the lower pO_2. They have increased lung volumes and also, in the case of Andean Indians, high hematocrits.[20] But no population of humans possesses a suite of adaptations that allows them to live for long periods of time with pO_2 levels less than half that at sea level.

Regulation

A FASCINATING question that remains unanswered is how the level of oxygen in Earth's atmosphere, sufficient to support the high metabolic rates of advanced life like ourselves, has for the past 200 million years been maintained at close to a pO_2 of 160. In a recent paper Sara Seager and William Bains concede, "There is no precise explanation for Earth's value of O_2 atmospheric abundance. In general, we have high atmospheric O_2 because of a combination of the presence of photosynthesizing life and the burial of some of that life as carbon, but what sets the precise level is not known."[21]

But whatever geophysical or biological mechanisms are involved in maintaining oxygen levels and atmospheric settings at close to current values, those mechanisms are entirely unrelated to their fitness for advanced terrestrial aerobes like ourselves. This counts therefore as another important example of environmental fitness for advanced aerobic terrestrial life even though there is no generally accepted causal explanation.

Spontaneous Combustion

THE NEED for very relatively high oxygen levels (pO_2 levels) in the atmosphere to support our own energy-hungry metabolism raises a question which intrigued novelist Arthur C. Clarke. Given the enormous quantities of energy released when reduced carbon compounds in vegetation or in animal bodies react with oxygen—"the reaction which powers the world"[22]—why don't we spontaneously combust at ambient temperatures, given the inherent thermodynamic energy of oxidations?[23]

And in the same vein, why don't fires rage catastrophically across all forested and vegetated regions, extinguishing all terrestrial life on earth save perhaps for a few primitive species in desert regions? Nick Lane raised the same question in his book *Oxygen*.[24] After all, as he points out, "the massive amount of heat released suggests that burning should proceed almost spontaneously... Regardless of whether we metabolize it or burn it, 125 grams [4 oz] of sugar (the amount required to make a sponge cake) will produce 1790 kilojoules... of energy: enough to boil 3 litres [6.3 US pints] of water or light a 100-watt bulb for 5 hours."[25]

The answer to the question lies in the unique "unreactive behavior" of carbon and oxygen atoms in the ambient temperature range. We have all experienced the curious reluctance of these two atoms to initiate a reaction in the difficulty of starting a campfire. The reaction (the fire) can only be initiated after some heat (from a match or fire lighter) is applied to the wood. The difficulty arises from the energy barriers one must overcome to initiate the reaction.[26] Although a considerable nuisance on a damp morning when attempting to start a campfire, these barriers are actually highly fortuitous, representing a crucial element of fitness in nature that allows us to live and breathe safely in an atmosphere highly enriched in oxygen.

Unreactive Carbon

The carbon atom's characteristic un-reactivity is witnessed in the everyday inertness of soot, graphite, and coal in the ambient temperature range. This inertness was underscored more than a century ago by Alfred Russel Wallace in *The World of Life* where he made the point by noting the custom of burning the ends of wooden posts so that when they are driven into the ground the coating of non-decaying carbon may preserve the inner wood.[27] Earlier still it was emphasized by William Prout (1834), where he commented, "Carbon in its elementary state is a very inert substance."[28] We now know that the reason for its relative lethargy resides in its unique electronic structure.[29]

Unreactive Oxygen

Oxygen, carbon's partner in combustion and respiration, is a highly reactive gas—one of the most reactive of all atoms at high temperatures, as witnessed in a forest fire. And yet it is curiously unreactive in the ambient temperature range. As Lane comments, "The fact that living things do not burst spontaneously into flame betrays an odd reluctance on the part of oxygen to react."[30] This reluctance stems from certain unique electronic characteristics of the oxygen atom. As M. J. Green and A. O. Hill explain, dioxygen (O_2), the common form of oxygen at ambient temperatures, is fairly unreactive due to an energy barrier that has to be overcome before this form of oxygen can be converted into various higher energy, more reactive species.[31]

The resulting low reactivity of O_2 in the ambient temperature range is crucial for life. Boulatov elaborates:

> The biosphere benefits greatly from this inertness of O_2 as it allows the existence of highly reduced organic matter in an atmosphere rich in a powerful oxidant. But such inertness also means that rapid aerobic oxidation will occur only if energy is put into the system to overcome the intrinsic kinetic barriers, or the reaction is catalyzed (i.e., the kinetic barriers are lowered by stabilizing otherwise high-energy intermediates). The energy of the reactants can be increased by raising the temperature... [the need for a match or fire lighter to initiate the camp fire] and indeed combustion is the most technologically important means for the utilization of the oxidizing potential of O_2. Combustion proceeds via [high-energy singlet] oxygen atoms, which are produced... at sufficiently high temperatures.[32]

But as Boulatov also notes, living organisms can't use this method due to their inability to tolerate such high temperatures. So instead they depend on metalloenzymes, containing transition metal atoms, "to lower the energies and the reactivities of partially reduced oxygen intermediates (such as O_2^-, H_2O_2) to accelerate, and to increase the efficiency of, O_2 reduction."

We are very fortunate that the reactivity of oxygen is greatly constrained at ambient temperatures. Otherwise, no reduced carbon compounds could persist for any length of time on the Earth's surface. There would be no photosynthesis, no plants, no forests, and certainly no humans.

In sum, we are able to live with an oxygen-enriched atmosphere— with a pO_2 of 160 mm Hg —to supply our energy needs, while at the same time avoiding Arthur C. Clarke's nightmare of spontaneous combustion, because of a crucial energy barrier which makes oxygen reluctant to react at ambient temperatures.

This benefits all advanced organisms deriving energy from oxidations, but particularly air-breathing aerobes living in a terrestrial environment immersed in an oxygen-enriched atmosphere. For aquatic aerobes—fish and invertebrates—the threat of spontaneous combustion is less: the concentration of oxygen is less in sea water than in the air (see comments above) and being immersed in water provides additional protection.

That oxygen is relatively unreactive at ambient temperatures serves two ends indispensable to life: (1) It protects air-breathing terrestrial organisms from spontaneous combustion at pO_2 levels sufficient to supply their energy needs. (2) It means the thermodynamic energy of oxidations can be exploited safely and controllably by all aerobes (aquatic and terrestrial) by way of, as we shall see later, the unique properties of the transition metals.

So in oxygen's relative un-reactivity we glimpse yet another element of fitness in nature of particular relevance for all aerobic life on Earth but particularly for terrestrial beings of our air-breathing physiological design. It is only because of its relative un-reactivity at ambient temperatures that the vast reservoir of oxygen in the atmosphere— generated intriguingly by the low solubility of oxygen itself—could be exploited by complex land-based life to supply our extravagant energy needs.

The Retardant

OXYGEN'S RELATIVELY unreactive nature at ambient temperatures has no influence on fire once it is initiated. Once started, as witnessed in a campfire, the heat of the fire generates sufficient heat to activate oxygen, overcoming the kinetic barrier so that the fire becomes a self-sustaining potentially dangerous chemical reaction. This self-sustaining chemical reaction is so effective at current oxygen levels that if it weren't for the fire-retarding effects of nitrogen in our atmosphere, wildfires might burn down the vegetation across whole continents.[33] Nitrogen's effectiveness as a fire retardant depends on two factors: (1) its relatively high heat capacity, which absorbs the heat of the flame, cooling the surrounding air, depressing the rate of oxygen activation, and reducing the speed at which the fire spreads; and (2) its high concentration in the atmosphere of 79 percent. Nitrogen's fire-retarding properties are quite dramatic. When the oxygen percentage in a mix of oxygen and nitrogen falls to about 15 percent (85 percent nitrogen), plant-based fuels will not stay alight,[34] and at less than 12 percent oxygen (88 percent nitrogen) all fire is extinguished.

The previous chapter noted nitrogen's retardant properties and other ways nitrogen is fit for life. The large amount of nitrogen in the atmosphere provides density and pressure to the atmosphere, which prevents the oceans from evaporating. Otherwise, oceanic evaporation would lead to increased amounts of water vapor in the atmosphere, and as water vapor is a potent greenhouse gas, this would trigger further atmospheric heating, threatening Earth with a runaway greenhouse.[35] And as also mentioned in the previous chapter, despite nitrogen's high concentration in the atmosphere, it poses no danger of a runaway greenhouse itself because, fortuitously, diatomic nitrogen (N_2) is not a greenhouse gas.

Atmospheric nitrogen is also the ultimate source of all the nitrogen atoms incorporated into living matter. Nitrogen is central to many biochemical processes and occurs in amino acids, the nucleotide bases of DNA and RNA, many vitamins, ATP, heme, and so forth. And as with oxygen, carbon, and many of the other key elements of life, there is no alternative candidate in the periodic table of elements that could substitute for nitrogen to serve these various ends. "No other atom," writes Arthur Needham, "can substitute for it except in very trivial ways."[36]

The key to nitrogen's stability in the presence of oxygen is the strength of the nitrogen–nitrogen triple bond,[37] which binds the two nitrogen atoms together in atmospheric nitrogen (N_2). This is considerably stronger than the oxygen–oxygen double bond in diatomic oxygen (O_2). And because of the strength of the N-N triple bond, atmospheric nitrogen is stable in the presence of oxygen, so the two gases can coexist in the atmosphere unchanged. A search through the properties of possible substitutes for nitrogen's role as fire retardant shows that very few gaseous substances are stable in the presence of oxygen, apart from nitrogen. Water vapor (H_2O) and carbon dioxide (CO_2) are two of the few other examples, but both have lower heat capacities than nitrogen and both are greenhouse gases, which can only be present in very small concentrations in the atmosphere lest they trigger a runaway greenhouse. The noble gases—neon, argon, and krypton—are also stable in the presence of oxygen, but they have lower heat capacities than nitrogen.

Without the fire-retarding properties of nitrogen, there could be no aerobic terrestrial organisms obtaining their oxygen by directly absorbing it from an atmosphere, much less terrestrial organisms able to control fire and develop technology. And I think that this almost certainly would apply throughout the universe. Wherever advanced air-breathing aerobic life thrives, it will be by permission of nitrogen.

Finally, a fascinating paper suggests that Earth may owe its nitrogen-rich atmosphere to another vital element of natural fitness—plate tectonics, the process of our planet's active geology by which large sections of Earth's crust gradually shift, forming and breaking up continents and subducting one plate beneath another over long periods of time. The authors of the *Nature* paper explain:

> Volatile elements stored in the mantles of terrestrial planets escape through volcanic degassing, and thereby influence planetary atmospheric evolution and habitability. Compared with the atmospheres of Venus and Mars, Earth's atmosphere is nitrogen-rich relative to primordial noble gas concentrations... We find that, under the relatively oxidized conditions of Earth's mantle wedges at convergent plate margins nitrogen is expected to exist predominantly as N_2 in fluids and, therefore, be degassed easily... We conclude that Earth's oxidized mantle wedge conditions—a result of subduction and hence plate tectonics—favour the development of a nitrogen-enriched atmosphere, relative to the primordial noble gases, whereas the atmospheres of Venus and Mars have less nitrogen because they lack plate tectonics.[38]

Earth's plate tectonics, in turn, depends on very exacting conditions being satisfied in the formation and evolution of the planet.[39]

Safe Combustion and Respiration

ONE MIGHT imagine that a high concentration of a powerful diluent (nitrogen) in the atmosphere to slow the spread of fire would depress the uptake of oxygen in the lungs. But in fact it has little effect. One might also imagine that the conditions which initiate fire and cause it to spread might influence the uptake of oxygen in the lungs, but the two processes are fundamentally distinct.

For the spread of flame, the critical factor is the heat capacity of the atmospheric gases. And in the case of our current atmosphere, it is the heat capacity of nitrogen (N_2), which makes up 79 percent of the atmosphere, that confers on the atmosphere its fire-retardant properties. "It was discovered that if the heat capacity of the atmosphere could be raised to ~50 cal/°C mole O_2, the atmosphere would not support combustion of any ordinary material," writes E. T. McHale. "Combustion depends on the feedback of energy on the flame zone to the unburned fuel in order to bring it to the combustion temperature. Inert gas diluents act as heat sinks for the combustion energy, cooling the flame and interfering with this feedback process and, at sufficiently high concentrations, quenching combustion."[40]

In contrast, for the uptake of oxygen in the lungs the heat capacity of nitrogen is irrelevant. The uptake of oxygen in the lungs is determined by the diffusion rate of O_2 into the blood from the alveolar air. This depends in turn on the pO_2 level in the lungs, which must be high enough to ensure diffusion of sufficient oxygen molecules from the air into the blood (250 ml or 10^{22} atoms per minute) across the alveolar wall to support our metabolic needs. As McHale explains, "The atmosphere plays a different role in sustaining life than in supporting combustion. The life support function requires a partial pressure... of oxygen sufficient to maintain the necessary oxygen tension in the blood. Diluent gases, if they are physiologically inert [like nitrogen], have only a minor effect on this process."[41]

And this difference has a remarkable consequence. "By selection of a proper additive it should be possible to prepare an atmosphere of high heat capacity that is also physiologically inert," McHale writes. "This would comprise a habitable atmosphere that would not support combustion."[42] And by "habitable" here he means one habitable specifically for air-breathers like ourselves. In other words, precisely because the factors which influence uptake in the lungs and the factors which influence the spread of fire are quite different, atmospheres that preclude fires but sustain oxygen uptake in the lungs are theoretically possible. Douglas Drysdale makes the same point in his book *Introduction to Fire Dynamics*:

> It is possible to create an atmosphere that will support life but not flame. If the thermal capacity of the atmosphere per mole of oxygen is increased to more than 275J/K (corresponding to 12% O_2 in N_2), the flame cannot exist under normal ambient conditions. A level of oxygen as low as 12% will not support normal human activity, but if this atmosphere is pressurized to 1.7 bar, the oxygen partial pressure will be increased to 160 mm Hg, equivalent to that in a normal atmosphere and therefore perfectly habitable—although incapable of supporting combustion.[43]

Figure 4.3. Edmund Hillary and Tenzing Norgay after completing the first ascent of Mount Everest on May 29, 1953.

We are so familiar with our atmosphere, which supports both fire and respiration, that we never consider that it is in any way unusual or that there might be atmospheres which support respiration but not fire and vice versa. But such atmospheres do exist. An example is the current atmosphere at altitudes above 8,000 meters in the Himalayas, described by climbers as a "death zone," since the atmosphere there cannot support terrestrial life for more than a short period of time, but can support fire. In John Hunt's *The Ascent of Everest*, New Zealander Edmund Hillary comments on the preparation at the South Col for the final push to the summit. He and Tibetan mountaineer Tenzing Norgay "started up our cooker and in a determined effort to prevent the weakness arising from dehydration we drank large quantities of lemon juice and sugar, and followed this with our last tin of sardines on biscuits… My boots were… frozen solid. Drastic measures were called for, so I cooked them over the fierce flame of the primus."[44]

Combustion also can occur at altitudes even higher than the summit of Everest, as demonstrated by passenger jetliners flying at 10,000 meters above sea level, propelled by the combustion of special petroleum products, at pO_2 levels far below those that can support human respiration for even a few minutes. The reason fire is sustained at such low pO_2 is mainly that the percentage of oxygen (21 percent) and the diluent (79 percent) remain the same. It is the percentage of oxygen in a nitrogen/oxygen mix that primarily determines whether an atmosphere will support fire, not its pO_2.

Unlike Earth's atmosphere in the higher altitudes, the atmosphere at sea level and up to about 5,400 meters does indeed support safe combustion and respiration, a fact of enormous consequence, since it allows us to both breathe and build fires. Remarkably, the basic parameters of the current atmosphere on Earth would appear to be the only set capable of supporting both safe combustion and respiration. These are a total atmospheric pressure range of 380–760 mm Hg, an oxygen pO_2 range of 80 to 160 mm Hg, and a percentage of oxygen and nitrogen of fairly close to 21 percent and 79 percent respectively. This is the only parameter set—the key formula as it were—that will generate a nitrogen/oxygen atmosphere supporting safe combustion and respiration.

Anyone can see this is so by selecting at random a parameter and changings its value and then working out the consequences. It soon becomes apparent that no significantly different alternative will work.

Suppose, for example, that combustion necessitated a percentage of oxygen in a nitrogen/oxygen mix of 50 percent, because the heat capacity of nitrogen gas was substantially increased above its current level. Then at an atmospheric pressure of 760 mm Hg this would give an oxygen partial pressure of close to 380 mm Hg. But this level would almost certainly be lethal for life because of the danger of oxygen free radicals (a topic discussed in the next chapter). Restoring the normal partial pressure of oxygen back to 160 mm Hg (and thereby reducing the danger of free radicals) and keeping the oxygen percentage at 50 percent by reducing the atmospheric pressure to, say, 380 mm Hg would not work either, as it would almost certainly lead to the evaporation of the oceans and eventually to a runaway greenhouse. Or suppose the partial pressure of O_2 in the atmosphere had to be 380 mm Hg for some reason to support human respiration (say, the solubility of oxygen was less and the rate of uptake in the lungs considerably reduced). Then assuming an atmospheric pressure of 760 mm Hg, the percentage of oxygen would be 50 percent and again fires would rage across the world.

Given that the atmospheric parameters which determine Earth's atmosphere at present must be very close to what they are to provide safe combustion and respiration over most of the habitable world, an obvious question arises. What mechanism determines these ideal parameters and how have they been maintained over the past several hundred million years at approximately current levels? The answer is, as in the case of the oxygen level, no one knows exactly. But whatever the mechanism involved in generating this ideal parameter set, it represents surely a compelling indication of the profound fitness in nature for terrestrial aerobic life on a planet like Earth.

Nature Expected Us

IN LIGHT of the evidence reviewed thus far in this book, can there remain any reasonable doubt that the basic order of nature anticipates terrestrial air breathers like ourselves? That nature was, in a sense, expecting us?

The facts speak for themselves. There is (1) the enormous quantities of energy released when oxygen combines with reduced carbon compounds, supplying advanced complex life forms like ourselves with copious quantities of metabolic energy, as well as the gift of fire and the means to send rockets into space. There is (2) oxygen's gaseous nature, essential for land-based life forms, allowing them to extract it from the atmosphere via lungs in the case of vertebrates, or via trachea in the case of insects. There is (3) the low solubility of oxygen, which prevents its loss to the oceans and preserves most of it in the atmosphere for the benefit of terrestrial life. There is (4) the light of the sun, which is just right for photochemistry, (5) the transparency of the atmosphere to visual light, and at the same time, (6) its absorption of a significant fraction of infrared radiation to warm the Earth into the ambient temperature range.

There is (7) the attenuation of oxygen's chemical vigor (the kinetic barrier) at ambient temperatures, which together with (8) the carbon atom's relatively low reactivity, prevents spontaneous combustion of reduced carbon compounds. There is (9) the fire-retarding influence of nitrogen—the only viable candidate gas for this role—which greatly slows down the spread of fire, rendering it controllable by fire makers like ourselves. There is the fact (10) that nitrogen and oxygen are not greenhouse gases and so do not directly affect the atmospheric temperature despite constituting 99 percent of the atmosphere. There is the fact (11) that because the atmosphere plays a very different role in sustaining respiration than in supporting fire, the presence of a fire retardant (nitrogen) in the atmosphere, which renders fire controllable, does not undermine the uptake of oxygen in the lungs.

There is (12) the fact that only trace amounts of ozone suffice to absorb dangerous ultraviolet radiation. And there are (13) the various absorption windows in the infrared region (including the crucial window between 8 and 14 nm), which prevent our atmosphere from turning our planet into a searing oven hostile to carbon-based life. And finally (14) there has been the maintenance of an oxygen partial pressure of about the current level, a level sufficient to support the energy demands of ourselves and other advanced terrestrial aerobes.

And this is by no means the only ensemble of fitness for aerobic terrestrial life. As we saw in Chapter 2, there is the ensemble of fitness that enables the hydrological cycle to deliver water and the essential elements of life for land-based life. As we saw in Chapter 3, there is the ensemble of fitness that enables the vital mechanism of photosynthesis, which provides us with the copious quantities of oxygen we need. And as we will see in future chapters, there are additional ensembles of fitness for advanced terrestrial organism like ourselves. Clearly nature went to great lengths to prepare for our arrival.

5. BREATHING

> The design of the human lung is impressive. The largest organ in our body is built with only about half a litre of tissue that separates roughly the same amount of blood from a large but varying air volume of several litres. And this tissue supports a very large gas exchange surface between air and blood—nearly the size of a tennis court— that must be ventilated and perfused with blood. This suggests that lung design is the result of bioengineering optimisation, making it a "good lung" serving its function, gas exchange, well and efficiently.
>
> — Ewald Weibel, "What Makes a Good Lung?"[1]

THERE IS NO PHYSIOLOGICAL PROCESS MORE FAMILIAR OR MORE vital than breathing. At rest we inhale and exhale about 500 ml of air every five seconds. In the course of a lifetime we take a total of nearly 500 million breaths. From the air we inhale into the lungs, we extract 250 ml of pure oxygen every minute at rest. This is carried in the red blood cells to the tissues, where it works its magic in the mitochondria, oxidizing the foodstuffs of the body and generating the metabolic energy we need to live. And every minute we exhale about the same volume of one of the waste products of those oxidations, the gas carbon dioxide.

Typically the main focus in evolutionary texts, and even in basic physiology texts describing respiration, is on the adaptative design of the lungs and how they might have emerged in the course of evolution. But what is hardly ever highlighted is that breathing is enabled by yet another unique ensemble of prior environmental fitness, without which

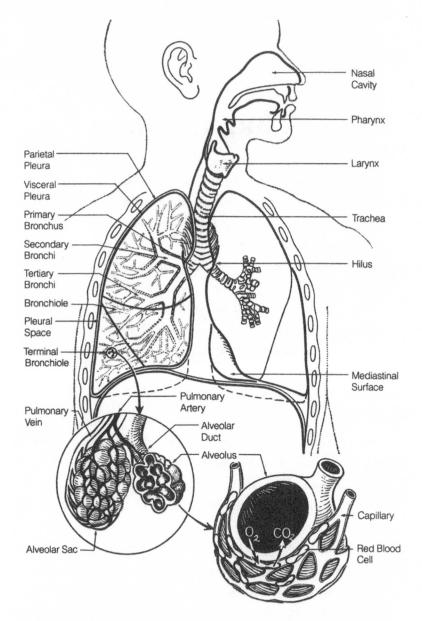

Figure 5.1. The human respiratory system. The trachea and major bronchi down to the terminal bronchi comprise the conducting zone. The terminal bronchioles and alveolar ducts and alveoli constitute the respiratory zone.

the adaptive design of the lungs could never have been actualized and no advanced aerobic organisms would grace the terrestrial domain.

In the text below, the focus is on the environmental fitness for our own mode of respiration, the mammalian bellows-type lungs. There are several other designs for air-breathing. Insects use a tracheal system of microtubes that permeate their tissues, which in many large insects involves adaptations for pumping the air through the trachea. Birds, unique among terrestrial vertebrates, pump air through a one-way "continuous through put lung" quite different from the mammalian bellows-type lung.[2] But many of the elements of environmental fitness which enable mammalian respiration enable all other types of air breathing as well.

Our Lungs

FROM FIRST principles, air breathing necessitates a lung, which is basically an invaginated gas exchanger. Such an invaginated organ for gaseous exchange is, as John Maina points out, an adaptive prerequisite "for water conservation on the desiccating *terra firma*."[3] As he continues:

> In animals, generally, no other organ is in more contact with an external environment than the gas exchanger. If, for example, the mature human lungs… were designed like gills (i.e., they were evaginated and exposed to air) even in a moderately desiccating environment, water loss would be about 500 L per day, a value about 1000 times more than normal loss….
>
> By the very nature of the lungs developing as invaginated organs, thus having a narrow entry/exit point to the ambient milieu, they can only be ventilated tidally, i.e. bidirectionally.[4]

Tidal ventilation, which moves air in and out of the lungs, is not the only mechanism involved in transporting air and its constituent gases, including oxygen and carbon dioxide. As James Butler and Akira Tsuda note, "The transport of oxygen and carbon dioxide in the gas phase from the ambient environment to and from the alveolar gas/blood interface is accomplished through the tracheobronchial tree and involves mechanisms of bulk or convective transport and diffusive net transport."[5]

Traditionally the respiratory tree has been divided into two regions. This first is a proximal region, the conducting zone in which bulk convection or ventilation moves the air in and out during breathing. The second is a distal region, the respiratory zone consisting of the terminal bronchioles and alveoli, where air movement is driven largely by diffusion. Butler and Tsuda elaborate:

> The classical description of gas transport from the ambient environment to the alveolar gas exchange region has been traditionally described in terms of a simple fractionation of the entire airway tree into a proximal region... where transport takes place cyclically by convection, and a distal region comprising a well-mixed alveolar space, where transport is effected by diffusion from gas to blood.... This picture is simple, but surprising in its ability to accurately characterize most features of gas transport in normal lungs.[6]

Our lungs are engineering marvels, but for them to function, air also has to be just so. We begin with its density.

The Density of Air

ONE VITAL element of prior fitness for ventilation is the low density of air. If its density had been significantly greater, the work involved in ventilation—pumping air in and out of the conducting zone—would be incommensurate with any type of respiratory system remotely comparable to our own. D. S. Minors explains how the work of breathing increases with the density of the air, noting that "the maximum voluntary ventilation is approximately proportional to the reciprocal of the square root of the density." As an equation that's ventilatory max = $K1/\sqrt{D}$, which means that at thirty meters deep in the water, the weight of air is about four times (4 bar) what it is at sea level. At such a weight, "the maximum voluntary ventilation of a man breathing compressed air is only 50 percent of that at sea level." Minors continues:

> Breathing compressed air also increases the work of breathing in another way: as the density of the air breathed increases, so flow of air in the airways becomes more turbulent, resulting in an increase in airway resistance. In addition, an increased density of gases hinders their intra-alveolar diffusion.

As a result of these factors, whereas maximum work capacity at sea level is normally limited by cardiovascular transport of oxygen, the limitations [posed by increased density] are largely ventilatory.[7]

With the weight of air as it is, the work of breathing at rest uses only 2 percent of total oxygen consumption.[8] But even with the very low density of air, during strenuous exercise the total work of breathing is considerable, with the respiratory musculature using up to 15 percent of the body's total oxygen consumption.[9] Even if the density of air were just, say, two or three times more than it is, given V max $\propto \sqrt{D}$, terrestrial aerobic life would be massively constrained. Strenuous activity would demand an intolerably high proportion of the body's total oxygen consumption. Human life would be a far more sedentary affair.

Atmospheric Pressure

BECAUSE THE weight and density of a gas varies directly with atmospheric pressure, it might seem that if the atmospheric pressure were less than its current level—760 mm Hg at sea level—then the work of breathing would also be less. But there are vital reasons unrelated to the work of breathing as to why the atmospheric pressure could not be much less than the current 760 mm Hg. For one thing, life on Earth—and on any habitable planet bearing carbon-based life—depends critically on the retention of liquid water, and this in turn, as mentioned in the previous chapter, depends on sufficient atmospheric pressure to prevent the oceans and other bodies of water from evaporating.[10] As pointed out, water vapor is a potent greenhouse gas, and increased water vapor in the atmosphere would increase atmospheric temperature, leading to more evaporation and eventually to a runaway greenhouse. When this happens, as the authors of a 2015 paper note, "Water vapor reaches the stratosphere, where it is easily photolyzed by UV radiation. Heating of the upper atmosphere by [UV] radiation can then drive a hydrodynamic wind that carries the hydrogen (and potentially some of the oxygen) to space, leading to the irreversible loss of a planet's surface water."[11]

As we saw previously, another reason atmospheric pressure could hardly be much less than the canonical 760 mm Hg is the need for considerable quantities of fire-retarding nitrogen gas in the atmosphere to prevent uncontrolled spread of fire—to keep oxygen, as it were, "on a leash."

Viscosity

THE VERY low viscosity of air compared with most other familiar substances is another factor that greatly facilitates breathing air (ventilation in the conducting zone).[12] The viscosity of air is fifty times less than that of water,[13] and many substances have viscosities many orders of magnitude greater than that of air or water. Indeed, the range of viscosities of common substances is far greater than the range of densities. Viscosity varies over many orders of magnitude, even among liquids. Thus, for instance, at 20°C honey is thousands of times more viscous than water.[14]

In the James Cameron science fiction movie *The Abyss*, the lead character played by Ed Harris uses a fluid-breathing system to allow him to dive deeper into the ocean than he'd otherwise be able to. However, liquid breathing as depicted in the film poses a number of challenges. As Maina points out, "Due to their physicochemical differences water breathing presents formidable problems to an air breather... Because of its higher viscosity, applying the same force, the flow rate of a liquid in the pulmonary passages should be slower than that of air. Moreover, to maintain equal flow rates, an air breather has to expend 60 times more energy to move [water]... than air."[15] And because of the increased density and viscosity of liquids, sustained liquid breathing requires mechanical ventilation.[16]

A gas substantially more viscous than air would pose a serious challenge. Indeed, if the viscosity of air were much increased, air breathing would be greatly constrained. The pressure (P) required to pump a substance through a tube rises with its viscosity. P = KV (where K is a constant), derived from the Hagen-Poiseuille equation.[17] And the flow of a substance through a tube is inversely proportional to its viscosity, i.e., Flow = K1/V.[18]

As it is, the viscosity of air is close to the lowest of any substance,[19] which makes it extremely well suited for efficient transport in the conducting zone.

Note that, counterintuitively, viscosity is largely unrelated to density. The density of mercury is thirteen times that of water but its viscosity is only marginally higher.[20] While honey is far more viscous than saline water, it is not as dense.[21] So the low density and low viscosity of air are, to a significant degree, independent parameters. But both are fit for air breathing.

The Compressibility of Air

ANOTHER ELEMENT of fitness for air breathing seldom highlighted is that air, being gaseous, is compressible. This additional property of gases further eases the work of breathing, particularly in our bellows-type lungs. During inhalation, the diaphragm and intercostal muscles contract, increasing lung volume and thereby decreasing the pressure in the lungs. This difference between the pressure in the lungs and that in the atmosphere draws air into the lungs, thus doing part of the work of inhalation.

Exhalation occurs when the diaphragm and intercostal muscles relax, decreasing lung volume and increasing the pressure in the lungs, and again, the differential pressure draws the air out of the lungs. A physiology textbook describes it this way: "Air flows down a pressure gradient, that is, air flows from an area of higher pressure to an area of lower pressure. It is this difference in pressures that drives pulmonary ventilation—the movement into and out of the lungs.... In order to breathe, we manipulate the volume of our lungs in order to change their pressure."[22]

Thus the compressibility of air enables the respiratory system to utilize differential pressure to do much of the work of breathing. Intriguingly, the very opposite property—the non-compressibility of liquid water—makes possible the circulatory system, a point we will return to in a later chapter.

An intriguing consequence of the mechanics of breathing is that while work is involved in inhalation, mainly due to the contraction of the diaphragm and intercostal muscles, much of the work of exhalation is accomplished by the spontaneous elastic recoil of the diaphragm and of the tiny air sac (alveoli) in the lungs, and requires little expenditure of energy. This recoil produces a positive pressure, which compresses the air in the lungs, increasing the pressure and causing the air to flow out of the lungs.[23]

Diffusion

AIR BREATHING also critically depends on the rapid diffusion of gases in air, because ventilation is ineffective in moving air in and out of the terminal bronchioles and alveoli. Gaseous transport in the periphery of the respiratory tree instead depends on the phenomenon of diffusion.[24]

If the diffusion rate of oxygen in air were the same as it is in water, the rate oxygen would reach the alveoli down the terminal branches of the respiratory tree would be massively decreased, and quite insufficient to supply air breathers with adequate amounts of oxygen to satisfy their metabolic needs. But in keeping with the fitness paradigm, the diffusion rate of gases in air is vastly higher than their diffusion rates in liquids. For example, the diffusion rate of oxygen in air is 8,000 times faster than in water,[25] precisely what is needed to transport oxygen efficiently in the terminal respiratory zone from the terminal bronchioles into the alveoli.

It is surely intriguing, and a further pointer to nature's fitness for air breathing, that the problem of supplying the terminal branches of the respiratory tree, which are "out of reach" of ventilation, is solved by the very high rates of diffusion of gases in air. Air is fit for ventilation in the conducting zone because of its low density and viscosity, and it's fit for gas transport in the terminal branches because of the high diffusion rates of gases in air. Often the evidence for the fitness paradigm resides in the more esoteric details of phenomena, and this is an example, one that has struck me ever since medical school as a remarkable example of fitness in nature for air breathing with lungs.

If one were to play the role of Plato's *Demiurge* and fabricate a substance with ideal properties for air breathing with lungs, we would end up inventing a gaseous substance with the precise physical properties of air, making sure it was fit for both ventilation in the conducting zone and for transport via diffusion in the respiratory zone where ventilation is inefficient.

It is only because air is light and of low viscosity that ventilation of the proximal bronchial passages (the conducting zone) is possible. And it is only because diffusion rates of gases in air are very rapid that oxygen and carbon dioxide can be transported efficiently and rapidly to and from the conducting zone via the terminal bronchioles and the alveoli at a rate sufficient to satisfy the metabolic needs of advanced high-energy aerobic organisms like ourselves.

In short, without the prior environmental fitness of air for air breathing, the uptake of the vast quantities of oxygen needed to support our high metabolic rate from an atmosphere would be impossible no matter how complex and sophisticated the adaptations involved.

Clearly air is fit for the transport of oxygen in both the proximal bronchi, in the conducting zone, and in the distal bronchioles and alveoli in the respiratory zone.

Fick's Law and the Interface

THE NEXT phase of breathing involves the respiratory gases oxygen and carbon dioxide diffusing across the gaseous exchange barrier between the air in the alveoli and the blood in the pulmonary capillaries.

The four factors which determine the rate of diffusion are given by Fick's law, which states that the rate of diffusion of a gas through a membrane is (1) inversely proportional to the thickness of the membrane, (2) directly proportional to the area of the membrane, (3) directly proportional to the difference in partial pressure across the membrane, and (4) directly proportional to its solubility in the membrane. Of these four factors the first two are the result of adaptation, while the third and fourth are like the density and viscosity of air, environmental givens and prior to the adaptive evolution of lungs.

Before reviewing how these environmental factors (factors 3 and 4) constrain the oxygen flux across the membrane, it is worth briefly considering the adaptive marvel of the lung for gaseous exchange, manifest in the extreme thinness of the gaseous exchange barrier (factor 1) and the extraordinary area (factor 2) of 130 meters square (three-fourths the area of a tennis court) over which it is spread and then packed into the modest volume of the lungs.[26]

Ultrathin

THE BASIC structural component of the exchange barrier, the alveolar membrane that surrounds each of the 500 million[27] alveoli in the lungs, is extraordinarily thin. In many places it is less than 0.2 microns across (0.2 thousandths of a millimeter, or 200 nm, about the thickness of the film encasing a soap bubble).[28] The alveolar membrane consists of a layer of extremely thin alveolar epithelial cells, which are so thin (25 nm) that an electron microscope was needed to prove that all alveoli are covered with an epithelial lining,[29] sitting atop a layer of the endothelial cells that line the pulmonary capillaries and a fused basement membrane between the two cell layers.

It is the extreme thinness of the membrane which enables highly efficient gas exchange.[30] In places between the epithelial cells and endothelial cells are bundles of collagen and elastin fibers. These confer the requisite tensile strength and elasticity to the ultrathin membrane,[31] enabling the alveoli to expand and relax with each inhalation and exhalation without quickly wearing out the membranes. Schmidt-Nielsen comments:

> The membrane that separates the air in the lungs from the blood must be thin so that oxygen can diffuse readily into the blood while carbon dioxide moves in the opposite direction. In the human lung much of the alveolar membrane where the gas exchange takes place is no more than 0.2 microns thick. What this means is difficult to imagine. The

thickness of a single page in this book is 50 microns; if one page could be sliced into 250 parallel layers, each would be the thickness of the alveolar membrane... [yet it's] strong enough to tolerate being stretched more than 20,000 times a day![32]

It is remarkable that a membrane so ultrathin, which must consist of at least two layers of cells—the epithelial cells of the alveoli walls and the endothelial cells lining the blood capillaries—can be bioengineered to sustain such repeated rounds of stretching and relaxation.

The extreme thinness of the alveolar membrane is clearly an adaptation and not an element of environmental fitness. However, because it is the basic physical properties of the relevant bio-materials which enable the construction of such an extremely thin membrane (less than 0.2 microns across), its design may be considered to be constrained by what are in effect prior elements of environmental fitness.

The alveolar membrane, it should be noted, is only one layer making up the gaseous exchange diffusion barrier between the air in the alveolus and the red blood cell. Oxygen and carbon dioxide also must diffuse through an aqueous layer made up of alveolar fluid (on the alveolar side) and a thin layer of plasma between the membrane and the red cell (on the capillary side). The total thickness of the barrier is no more than about 2 microns.[33]

Area

THE THINNESS of the alveolar membrane is even more remarkable considering the extraordinary area over which it is spread. Covering the surface of all 500 million alveoli, it makes up a total area of about 130 square meters. This is equal to about three-fourths of a singles tennis court.[34] As Ewald Weibel points out in the journal *Swiss Medical Weekly*:

> The two key design features of "good lung" are hence a very large surface area and a very thin tissue barrier, features that are precarious... [This raises] two questions of physiological significance: (1) how is it possible to accommodate this surface within the limited space of the chest cavity and still allow for efficient ventilation and perfusion; and (2) how is it possible to support, maintain, and stabilise a surface area

of the size of a tennis court with so little tissue?... the bioengineering problems to be solved by design are (a) how to build a sprinkler system to supply a few hundred million gas exchange units with O_2-rich air and with blood, and (b) how to fold up a sheet of 130 m² to fit into a space of 5 liters—equivalent to packing a letter into a thimble—while ensuring precise connections of the sprinkler system to the surface units [that is, to the bronchioles on one side and the blood capillaries on the other].[35]

The design of the lung is certainly, as Weibel puts it, "impressive."[36] That such an ultrathin membrane, having a total mass of about 500 grams, can be stretched robustly over such a huge area—one hundred times the surface area of the human body[37]—and crumpled up into the 5–6 liter volume of the lungs, is an extraordinary feat of bioengineering. It is hard to imagine how any greater area of such an ultrathin membrane could be compacted into the lungs. In short, the lungs are an adaptive wonder representing the result of, as Weibel puts it, "bioengineering optimisation."[38]

Partial Pressure

The third factor determining the rate of diffusion given by Fick's law is the difference in partial pressure across the membrane. According to Fick's law, the diffusion of a gas across a diffusion barrier is directly proportional to the difference in partial pressure across the barrier.[39] In the case of gaseous exchange in the lungs, it is only because there is a substantial difference in partial pressure across the alveolar barrier that oxygen diffuses from the alveoli into the blood. And this difference is determined by a number of basic environmental factors prior to and having nothing to do with the adaptive wonder of the lung. Primarily and perhaps the most important of all is the atmospheric pO_2 of 160 mm Hg—the pO_2 in the air drawn into the lungs. As mentioned in the previous chapter, just how this figure has been maintained over hundreds of millions of years is not fully understood, but unless the atmosphere contained a relatively high concentration of oxygen approaching that of our current atmosphere, human aerobic respiration (and the respiration of all other aerobic organisms—both terrestrial and aquatic) would be severely compromised.

One inevitable consequence of the design of our lungs as invaginated gaseous exchangers is as air is drawn from the mouth or nose down the respiratory tract to the alveoli, the partial pressure of oxygen (pO_2) declines. Whereas the pO_2 in the atmosphere at sea level is close to 160 mm Hg, by the time the air reaches the alveoli, its partial pressure has decreased by 55 mm Hg to about 105 mm Hg. This decrease is inevitable because of two factors: the humidification of the air as it is drawn through the respiratory tract and the presence of substantial quantities of CO_2 gas in the alveoli.[40] About 11 mm Hg of the decline—to a pO_2 of 149 mm Hg—is due to water vapor, and about 44 mm Hg —to a pO_2 of 105 in the alveoli—is due to the presence of CO_2.[41]

The presence of water vapor in the respiratory tract is an unavoidable consequence of the fact that water is the matrix of the cell and the medium of the blood, and makes up as much as 60 percent of the mass of the body in an adult human. It is a crucial element of fitness for air breathing that the vapor pressure of water at body temperature is relatively low and only reduces the pO_2 in the respiratory tract a small degree. And again, as in so many other instances, such as the fire-retarding property of nitrogen, or the delivery of water to the land via the hydrological cycle, this is another element of fitness specifically for air breathing, one quite irrelevant to a water-breathing organism breathing through gills.

The saturated vapor pressure of water (pH_2O) at ambient temperatures is quite low. At 20°C it is 17 mm Hg and at 37°C is still only a modest 47 mm Hg. But at temperatures greater than body temperature, it rapidly rises. At 50°C it is 92 mm Hg; at 60°C, 150 mm Hg; and at 80°C, 355 mm Hg.[42] Calculation shows that at a water vapor pressure of, say, 355 mm Hg, the pO_2 in the respiratory tract would fall from 149 mm Hg (the normal value) to about 85 mm Hg[43] and the pO_2 in the alveoli to about 35 mm Hg,[44] to the same level near the summit of mount Everest and far too low to sustain human life for any prolonged period.[45]

Fortunately, the saturated vapor pressure of water drops precipitously from 80°C down to the human body temperature of 37°C (98.6°F), to seven times less than it is at 80°C.[46] And because the vapor pressure of water at body temperature is a modest 47 mm Hg, this only reduces the pO_2 in the respiratory tract by 11 mm Hg, from 160 mm Hg to 149 mm Hg.

In passing, it is worth noting that the relatively low vapor pressure of water at ambient temperatures is a vital element of fitness for all life on Earth, not just air breathing organisms like ourselves. It means that, despite the fact that water is a potent greenhouse gas, the amount of water vapor in the atmosphere at Earth's mean surface temperature (about 15°C)[47] is insufficient to provoke a runaway greenhouse, which would lead to the loss of the oceans and the ultimate dehydration of the planet. But the amount is quite sufficient to enable the hydrological cycle, which provides the vital water for land-based life.

Another gas present in the respiratory tract is carbon dioxide, an inevitable chemical product of the oxidation of the reduced carbon compounds in the tissues, the reaction which, as we have seen, provides the metabolic energy for all advanced life on Earth.

$$CH + O_2 \rightarrow CO_2 + \text{water} + \text{energy (ATP and heat)}$$

As a result of this reaction, CO_2 is continuously generated in the tissues and carried in venous blood to the lungs, where it diffuses across the interface into the alveoli. From there it diffuses up the terminal bronchi in the respiratory zone into the conducting zone from which it eventually exits the lungs via ventilation.

Its gaseous nature at ambient temperatures is another vital element of fitness for air breathing, as it means it can be readily excreted in the lungs of terrestrial organisms via the respiratory tract—the same route through which oxygen is absorbed. If it were not a soluble gas, its excretion would poses herculean physiological problems, as Lawrence Henderson pointed out.[48]

In addition to the gaseous nature of carbon dioxide, another important element of fitness for air breathing is that the ratio of molecules of oxygen (O_2) consumed to carbon dioxide (CO_2) generated (what is known as the respiratory quotient) is close to 1. This means that every molecule of oxygen absorbed in the lungs and used in the mitochondria yields approximately one molecule of CO_2, depending on the type of foodstuff metabolized.[49] And it is this ratio (close to 1:1) that results in the pCO_2 in the alveoli of about 44 mm Hg.[50] If it were, say, 1:2, the body would be flooded with excess CO_2 waste and the pCO_2 would rise well above 44 mm Hg, leading to a further decrease in the alveolar partial pressure of oxygen.[51] As it is, although forced hyperventilation can lower the pCO_2 in the lungs below pCO_2 of 44 down to nearly 10 mm Hg, as reported in some studies, the effort required rapidly causes respiratory fatigue and cannot be sustained.[52]

Together, the pH_2O (11 mm Hg) and pCO_2 (44 mm Hg) reduce the pO_2 in the alveoli to 105 mm Hg.[53] This pO_2 is 60 mm Hg more than the pO_2 in the mixed venous blood entering the lungs (40 mm Hg);[54] and although this is a substantial reduction, the difference is more than sufficient to drive oxygen across the interface and fully oxygenate the mixed venous blood passing through the lungs (in humans and in all other mammals) so that the pO_2 in the blood leaving the lungs is 100 mm Hg, and the hemoglobin in the red blood cells is fully saturated with oxygen[55] even during strenuous exercise, when the extraordinary amount of five liters of oxygen is uploaded every minute.[56]

If the pO_2 in inspired air were substantially less than 160 mm Hg, or if the pH_2O and pCO_2 in the respiratory tract were substantially greater, the only way to rescue the necessary rate of oxygen diffusion across the membrane into the blood would be to further decrease the width of the alveolar membrane or further increase the area of the interface.[57] But from all the evidence available, this is not feasible. As far as the alveolar membrane is concerned, it could hardly be thinner, and the layer of water on both sides could hardly be reduced. And the area of the interface, assuming that the packing is maximal, could not be much greater.[58]

Solubility

As the fourth of the four factors delineated in Fick's law states, the rate of diffusion of a gas through a liquid membrane is directly proportional to another prior environmental given, its solubility.[59] And as the exchange barrier through which oxygen must diffuse is largely composed of water—alveolar fluid, intracellular water, blood plasma, etc.—its rate of diffusion will be approximately proportional to its solubility in water.

The solubility of gases in water varies over a range of 300,000, from ammonia (one of the most soluble) to hydrogen (one of the least soluble).[60] The solubility of oxygen is relatively low compared with many gases.[61] For example, it's about twenty-two times less soluble than carbon dioxide when in contact with the blood plasma of the human body.[62]

Oxygen's solubility is low, but not too low. The fact that it is soluble to a degree is an environmental given which enables it to diffuse across the interface. If it were insoluble, or as relatively insoluble as hydrogen (twenty times less soluble than oxygen), then both water breathing and air breathing would be impossible. The diffusion rates of oxygen in the mammalian lungs would be so low that the area of the lungs required to rescue the canonical flux of 250 ml per minute would be not a tennis court but a soccer field, requiring a lung volume many times too large to fit into the human body even with the extraordinary degree of compaction noted above. Moreover the water in the oceans would be far too anoxic to support aerobic metabolism in large aquatic organisms breathing through gills.

Even as it is, oxygen's relatively low solubility means that most of the oxygen in all vertebrates must be carried from the lungs to the tissues in the red blood cells by the oxygen-carrying molecule hemoglobin. In mammalian blood each 100 ml leaving the lungs contains about 20 ml of oxygen,[63] sixty times more than could be transported in simple solution, nearly all of which is bound to hemoglobin. In invertebrates, including the Cephalopods and the Crustacea, the oxygen is transported in the blood by the blue-colored protein hemocyanin.

Since medical school I often have wondered what end the low solubility of oxygen might serve. The flux of oxygen across the 2-micron aqueous diffusion barrier and into the red cells would be far more rapid if the solubility of oxygen in water were many times greater (like that of many gases, such as carbon dioxide). Then not only would it cross the membrane much faster, it would also dissolve in sufficient quantities to supply the body with oxygen without any need for an oxygen carrier like hemoglobin.[64] But over the years in which I have been working in the area of fitness, I have become aware of several elements of fitness stemming from oxygen's low solubility.

First, as mentioned in the previous chapter, because of its low solubility, only a small proportion of the world's oxygen dissolves in the oceans, leaving the majority in the atmosphere[65] where it can provide the high atmospheric pO_2 levels essential to support the high demands for oxygen of advanced terrestrial organisms.

Secondly, oxygen's low solubility may protect aerobic organisms from increased levels of potentially dangerous oxygen free radicals in the body. Oxygen free radicals—such as peroxide (H_2O_2), superoxide radical (O_2^-), and hydroxide ion (OH^-)—are reactive oxygen species generated in the body as oxygen undergoes a series of reductions (by the successive addition of electrons) catalyzed by transition metal atoms.[66] These cause diverse types of oxidative damage, including lipid peroxidation and DNA mutagenesis, and are implicated in aging and disease even at current pO_2 levels.[67] The more oxygen in solution, the greater the concentration of potentially dangerous free radicals in the body. If oxygen were sixty times more soluble (and thereby dispensing with the need for hemoglobin), their concentration in the body would be far higher. Moreover, free radicals are also present in sea water,[68] and if oxygen were many times more soluble, their concentration in the oceans and natural bodies of water would also be increased, with likely deleterious effects on aquatic life.

Given the potential danger of free radicals, hemoglobin is a very clever adaptation, enabling the transport of sixty times more oxygen than could be carried in simple solution, safely caged in a reversible chemical association with the iron in hemoglobin and kept from activation and the formation of dangerous free radicals.

Just how dangerous free radicals are at current levels in the body remains a matter for debate. Cells defend themselves effectively against these reactive products by means of antioxidant metalloenzymes, including the superoxide dismutases, catalases, and peroxidases.[69] These enzymes are in some instances extraordinarily efficient[70] and are an essential prerequisite for the use of oxygen by living cells.

Recently some researchers have argued that free radical generation at current levels of pO_2 in the body may be beneficial, even essential. There is evidence, for example, that they prolong the life of certain organisms and are even implicated in eliminating tumor cells.[71] Moreover some creatures live for hundreds of years,[72] some trees may live for several thousand years,[73] and some organisms, like the flatworm and certain species of jellyfish, are in effect immortal.[74] This demonstrates that, at least in certain cases, the damage inflicted by free radicals on extant life, protected by a battery of antioxidant enzymes, has little to no long-term deleterious effect.

So, many organisms can live and even thrive at the current pO_2 of 160 mm Hg. The way the literature is trending it seems that it may turn out that a normal level of oxygen free radicals are not just beneficial but actually essential for health, and that only higher levels than normally generated in the body—which would follow from higher atmospheric pO_2 or greater solubility of oxygen in water—will prove damaging. My sense is that this will prove to be the case.

It would seem that oxygen's relatively low solubility, like its "reluctance to react" in the ambient temperature range, is part of the oxygen paradox—that life needs the energy of oxidations provided by this highly reactive and dangerous element, but can only utilize oxidations when oxygen is on a leash, restrained by its unique spin restriction, caged and safe in the body by association with transition metals (as in hemoglobin), tamed by fire-retarding nitrogen in the atmosphere, and so forth.

Moreover, despite oxygen's relatively low solubility and diffusion rate compared with many other gases, its diffusion rate is perfectly sufficient given the adaptive wonder of the lung to fully oxygenate the 70 ml of blood pumped by the right ventricle through the pulmonary capillaries every second so that the pO_2 of the blood leaving the lungs is virtually the same as that in the alveoli.[75] So in effect the low solubility and relatively low diffusion rate pose no limit on the uptake of oxygen given the current adaptive design of the mammalian lung and the protein wonder that is hemoglobin.

The Right Volume

THE ULTRATHIN alveolar membrane in our lungs would cover most of a tennis court if spread out. As discussed above, that this membrane is packed into the volume of an adult human lung of about 5–6 liters[76] is an adaptive marvel. It also indicates another element of fitness for air breathing in organisms of our basic mammalian body plan: given the physical constants for gaseous exchange in the lungs—the pO_2 of the atmosphere, the pH_2O of water vapor at 37°C, the pCO_2 in the respiratory tract, the respiratory quotient of 0.8, the diffusion rate of oxygen in water, and so forth—the functional volume of lung parenchyma necessary to ensure the canonical uptake of the 250 ml of oxygen in an adult human (at rest) is no more than about 5–6 liters, occupying a volume of roughly 8 to 10 percent of the total volume of the body of a 70 kilogram (150 pound) adult,[77] fit for placement within the human body.

Assuming that the current compaction of alveoli in the lungs is just about as dense as it could be, it is surely a very significant element of fitness that a lung of volume of about 5–6 liters is sufficient to provide the gaseous exchange area necessary to fully oxygenate the blood and supply the energy demands of an adult human. The volume could hardly be much larger or the lungs would then take up an unacceptably large proportion of the body's volume.

One can easily imagine the problem of fitting into the human body a "super lung" of, say, twenty liters, thereby occupying nearly a third of the volume of the body, and the need to reengineer a far larger rib cage to encase such a monstrous lung.

Studies of high-altitude populations, such as Tibetans and Andean Indians, indicate that while at high altitudes increased lung volume and area for gaseous exchange would be adaptive, only relatively small increases in lung volume have occurred even after hundreds of generations of living at high altitudes.[78]

Moreover, any increase in lung volume and number of alveoli (and an increase in membrane area) would be of little utility without an equivalent increase in blood volume to perfuse the oversized lungs to prevent what is termed "ventilation perfusion mismatch," where many alveoli would not receive an appropriate blood supply for gaseous exchange.[79] There would be little point in having a gaseous exchange membrane the area of two or more tennis courts, and a lung of 20–30 liters, without a massive increase in the volume of blood. As it is, the blood vessels and blood take approximately another 10 percent of the volume of an adult human,[80] so there is little scope for an increased volume of blood to compensate for a theoretical increased lung volume. And again, the volume taken up by the blood in all mammals is about the same, about 10 percent of body volume.[81] (The proportion of the body's mass taken up by the circulatory system is discussed further in Chapter 6, and the remarkable fitness of the proportion of the body's volume taken up by other major organ systems is discussed in Chapter 9.)

It is surely an arresting fact that a functional lung volume which enables the uptake from the air of sufficient oxygen to satisfy our metabolic demands is commensurate with our anatomical body plan. This functional volume is determined by the adaptive fine tuning of lung function—discussed above—and by a set of physical determinants: the density and diffusion rates of gases in the air, the pO_2 in the atmosphere and in the alveoli, the diffusion rate of oxygen across the membrane, the ratio of about one molecule of CO_2 produced for every one molecule of oxygen used, the pH_2O at 37°C, the pCO_2 in the lungs, and so forth.

From First Principles

IT WOULD seem from the evidence reviewed in this and the previous chapter that all advanced large aerobic terrestrial carbon-based organisms of about our size anywhere in the cosmos will need, like us, to upload oxygen from an oxygen-rich atmosphere with close to pO_2 level of 160 mm Hg.[82] They will breathe under blue skies[83] from an atmosphere composed of a mix close to 20 percent O_2 and 80 percent N_2 (a mix which enables respiration and the controlled use of fire). And their lungs—what Maina refers to as "invaginated respiratory organs... a requisite for water conservation on the desiccating terra firma" if tidally ventilated—will contain a dichotomous branching system leading from a trachea via a series of large bronchi through a conducting zone to a respiratory zone consisting of narrow bronchioles leading to a vast number of small air sacs or "alveoli" where gaseous exchange occurs. There will be water vapor and carbon dioxide in their lungs. Additionally, given the solubility of oxygen and its diffusion rate in aqueous fluids to upload 250 ml of oxygen per minute, they will need an ultrathin membrane (about 0.2 microns thick) spread out over an area near to that of a tennis court. And their blood will be spread over their alveoli in an exceedingly thin film some four microns across. In short, their lungs will be, in several important ways, like our own.

A skeptic may be tempted to dismiss such speculation as unimaginative and hopelessly anthropocentric. But given the complete absence in the scientific literature of any well-worked-out models describing alternative non-carbon-based biochemistries and descriptions of an alternative type of cell, or any carefully worked out alternatives to oxidations to supply advanced carbon-based life forms with the copious amount of metabolic energy needed to support their extravagant active lifestyles, it is surely the biological mainstream which might be accused of being unimaginative for failing to consider another paradigm, for the failure of researchers to describe any alternative scenarios is surely a smoking gun, providing compelling support for the anthropocentric paradigm.

Just Right

THIS CHAPTER has added another ensemble of fitness to those discussed in previous chapters. This new ensemble includes: (1) the fitness of the low density and viscosity of air for ventilation in the conducting zone in the lungs, which greatly decreases the work of breathing; (2) the fitness of the high rates of diffusion of gases, which enables gaseous transport in the respiratory zone; (3) the fitness of the high pO_2 in the atmosphere; (4) the fitness of relatively low pH_2O in the respiratory tract at ambient temperatures; (5) the fitness of the gaseous nature of carbon dioxide; (6) the fitness of the ratio of O_2 inhaled to CO_2 exhaled, about 1:1, which limits the pCO_2 in the blood and respiratory tract to about 40 mm Hg; (7) the thinness of the alveolar membrane, which allows for rapid gaseous exchange; and finally (8) the functional volume of the lungs of about 10 percent of the volume of the mammalian body to provide suf-ficient alveolar area, given the prior physical constraints for adequate oxygen uptake.

What makes the ensemble so compelling are the exquisite teleo-logical details. For example, where gas transport via ventilation cannot reach, the rapid diffusion of gases comes to the rescue. Or there's the fact that, thanks to multiple environmental givens, our lungs need not take up an inordinate proportion of the mammalian body in order to supply sufficient oxygen.

In short, nature is fit for air breathing and for complex aerobic organisms like ourselves to upload sufficient oxygen to satisfy our energy demands. That the necessary rate of uptake is at all possible depends on many physical parameters being just right.

Moreover, as we shall see in Chapter 6, there is another essential ensemble of fitness which enables the existence of a circulatory system able to transport the vital oxygen from the lungs to the tissues, without which all the many elements of fitness for living with oxygen and for air breathing reviewed thus far would be to no avail.

6. CIRCULATION

So the heart is the center of life, the sun of the Microcosm,
as the sun itself might be called the heart of the world.
The blood is moved, invigorated, and kept from decaying
by the power and pulse of the heart. It is that intimate
shrine whose function is the nourishing and warming
of the whole body, the basis and secret of all life. But of
these matters we may speculate more appropriately on
considering the final causes of this motion.

—William Harvey, *De Motu Cordis* (1628)[1]

ALL BIG, complex organisms, including ourselves, require a circulatory
system. This is due to a fundamental physical constraint—diffusion's
inefficiency as a transport mechanism over distances greater than a
fraction of a millimeter. Diffusion is fast over very short distances, but
grows progressively slower the farther it has to reach. Just how dramatically slower was illustrated by physiologist Knut Schmidt-Nielsen. He
estimated that oxygen diffusing into the tissues will achieve an average
diffusion distance of one micron in one ten-thousandth of a second, ten
microns (approximately one cell diameter) in one-hundredth of a second,
one thousand microns (one millimeter) in one hundred seconds, and ten
millimeters (one centimeter) in three hours.[2]

The ability of small molecules to diffuse rapidly over short distances
in aqueous solutions explains why unicellular microorganisms, including bacteria and protozoa, and even very small multicellular organisms,
can obtain their nutrients and discard their waste simply by diffusion,
without the need for a circulatory system. Diffusion rates in water are

also fast enough over the short distances in play in intracellular metabolic processes. But diffusion's slowness over distances of more than a few cell diameters explains why no organism more than a few millimeters thick acquires and disposes of its metabolites by diffusion alone. This constraint necessitates some sort of circulatory system.

In most invertebrates the circulatory system is what is called *open* and the "circulation" involves basically pumping body fluids that carry oxygen and other dissolved nutrients around body cavities. In the case of insects the circulatory system is also open, but oxygen is provided not via circulation but, as mentioned in the previous chapter, by a series of tiny tubes, trachea, which permeate the body tissues and even individual cells, bringing oxygen directly to the metabolizing tissues.[3]

But many large complex multicellular organisms—including invertebrates, such as the squid and octopus, and all vertebrates, including mammals and birds—possess a closed circulatory system consisting of a vascular tree which undergoes a series of bifurcations leading from a pumping device (a heart) via increasingly smaller vessels and terminating in billions of tiny capillaries about 5–10 microns wide (in mammals) and about one millimeter long, which permeate the tissues of the body.[4]

The functioning of the various types of circulatory systems depends not only on a variety of clever adaptations (including pumping mechanisms such as hearts and contractile vessels) but on a crucial element of prior environmental fitness—a fluid with the right ensemble of properties to serve as the basic medium of the circulatory system. And that medium, as it happens, is the same one—water—ideally suited for another very different circulatory system, the grand hydrological cycle discussed in Chapter 2, a cycle which delivers water to the land, making terrestrial life possible.

The Fitness of Water

FROM FIRST principles, the medium of any functioning circulatory system in any type of complex multicellular organism resembling ourselves would have to possess a unique ensemble of fitness: (1) incompressibility,

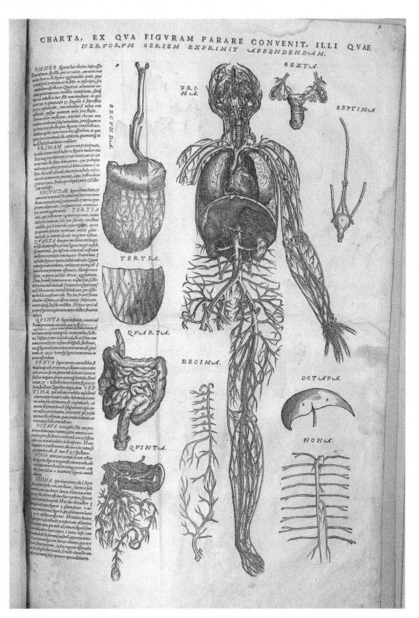

Figure 6.1. Blood vessels, heart, and lungs. From Andreas Vesalius's *De Humani Corporis Fabrica* (1543).

since compressible substances cannot be easily pumped; (2) a relatively low density and weight, commensurate with the power output of biological pumps; (3) the ability to dissolve oxygen, many different nutrients, and the waste products of metabolism; and (4) a relatively low viscosity, since viscous materials cannot be pumped through tiny capillaries.

And here again nature has obliged. All the necessary elements of fitness for the primary medium of a circulatory system, a vital necessity for all large multicellular organisms, are wonderfully satisfied in that most familiar of all fluids, water. In the text below we consider primarily the fitness of water to serve as the medium of the closed circulatory system of higher vertebrates including ourselves.

Water's Incompressibility

Like all liquids, water is essentially incompressible,[5] a vital physical property of any medium which has to be pumped at pressure through a vascular tree. One can well imagine the problem of pumping a compressible substance. The blood would never leave the heart! In the unicellular realm, where organisms have no use for a circulatory system, water's incompressibility might be of little consequence, but for organisms that require a circulatory system, its incompressibility is a crucial element of fitness.

Water's Deformability

The medium of a circulatory system must not only be incompressible (and capable of being pumped) it must also be readily deformable and able to conform to the various shapes through which it is pumped. In a mammal these include the left ventricle, the aorta, capillaries, the venous sinuses in the spleen, and so forth. Only one of the three familiar states of matter (solid, liquid, and gas)—the liquid state—will satisfy these two criteria. A solid is incompressible (or nearly so) but is not readily deformable. A gas is highly deformable and able to conform to the shape of any container, but it is also highly compressible. The fact that liquids are both incompressible and deformable may be considered therefore as the first two prime elements of prior environmental fitness in nature for circulatory systems. Moreover, if the liquid state of matter did not exist in the ambient conditions on Earth, there would be no circulatory systems. No complex multicellular organisms would be possible, and the only form of carbon-based life would be primitive unicells.

In addition to these two prime generic properties of all fluids, water also possesses specific elements of fitness to serve as the major constituent of the blood.

Water as Solvent

We saw in Chapter 2 that water's immense powers as a solvent play a vital role in the weathering and erosion of the rocks, a process that delivers the essential minerals for life onto the land. Those same powers play a crucial role in our bodies, since the basic medium of any circulatory system must be an excellent solvent to transport around the body the vast inventory of chemical compounds involved in the various metabolic processes upon which the life of complex living things depend.

"Liquid water is such a good solvent," Alok Jha remarks, that it's "almost impossible to find naturally occurring pure samples and even producing it in the rarefied environment of the laboratory is difficult." The reason, as he explains: "Almost every known chemical compound will dissolve in water to a small (but detectable) extent."[6]

No other liquid comes close. Virtually all organic substances that carry either an ionic charge or contain polar regions—which include the great majority of all organic compounds in the cell—are readily dissolved in water. Even large molecules such as proteins can dissolve in water if they have ionic or polar regions on their surface.[7]

In addition to water's generic fitness as a solvent, its fitness is also manifest in the very different solubilities in water of two vital compounds—oxygen and carbon dioxide. We have already seen in Chapter 4 that oxygen's relatively low solubility greatly lowers the amount of oxygen in the oceans, allowing it to accumulate in vast quantities in the atmosphere for the benefit of terrestrial air-breathings organisms. The solubility of carbon dioxide in water is many times more than that of oxygen[8] and results in an absorption coefficient far higher than that of oxygen, close to one at ambient temperatures. This means that the volume of CO_2 in one liter of an atmosphere in contact with a body of water or blood is the same as in one liter of water. This is a very important element of fitness for all life on Earth, not just for aerobic organisms. Lawrence Henderson explained why in his classic *The Fitness of the Environment*:

> Carbon dioxide has an absorption coefficient nearly equal to one... Hence, when water is in contact with air, and equilibrium has been established, the amount of free carbonic acid in the water is almost exactly equal to the amount in the air... Thus the waters can never wash carbonic acid out of the air, nor the air keep it from the waters. It is the one substance which thus, in considerable quantities relative to its total amount, everywhere accompanies water...
>
> Carbonic acid thus possesses the first great qualification of a food: its occurrence is universal and its mobility a maximum. This is due to the fact that its absorption coefficient [in water] is on the average approximately one, the most favourable value.[9]

Water's Density

Water, like many liquids, is relatively light compared with solids such as metals or minerals. Salt is over twice as dense. Iron is eight times as dense. Mercury, thirteen times as dense. And tungsten, more than nineteen times as dense.[10] One can well imagine the difficulty of pumping blood had water been even twice or three times the density and weight. The five liters of blood in the average human would have weighed not five kilos but ten or fifteen. Probably no circulatory system would have been possible. As Steven Vogel points out, even as things stand, empowering the circulatory system consumes about 10 percent of the resting energy of the body.[11]

Water's Viscosity

Another property of water with a critical bearing on the functioning and design of the circulatory system is its viscosity. Water's viscosity determines two important parameters. One is the diffusion rate of nutrients and solutes (such as oxygen) in aqueous solutions (such as blood). The second is viscous drag. This is the resistance experienced in moving an object in a liquid. (Think of moving a spoon through honey.) If water's viscosity were much increased the hydrostatic pressure required to pump the blood through the circulatory system would soon become prohibitive. This is because the pressure (P) needed to pump a fluid through a pipe is directly related to its viscosity (V). In other words, pressure rises with viscosity—$P=KV$, where K is a constant (derived from the Hagen–Poiseuille equation).[12]

As it is, the pressure at the arterial end of a human capillary is 35 mm Hg, which is considerable. This relatively high pressure is needed to force the blood through the capillaries. This would have to be increased dramatically if water's viscosity were several times higher. But such an increase would be incommensurate with any sort of biological pump. As mentioned above, empowering the circulatory system consumes about 10 percent of the resting energy of the body. Consequently, as with density, only limited increases in viscosity could be tolerated. One can easily imagine the insurmountable energetic challenge if water had the viscosity of honey, thousands of times greater than water,[13] or even of olive oil, more than eighty times greater.[14]

A substantially greater viscosity would have another deleterious consequence besides an increase in energetic costs. As we saw, a fluid's viscosity determines diffusion rates of solutes in that fluid. Diffusion rates are inversely related to viscosity, i.e., $D=K/V$, where D is the rate of diffusion, K is a constant, and V is viscosity. The greater the viscosity, the lower the diffusion rate.

As things are, a molecule attains an average diffusion distance of 100 microns—the thickness of ten cells—in one second. But if the viscosity of water were similar to honey's, it would take an hour to cover the same diffusion distance and minutes to traverse one cell. As things stand, increasing the viscosity of water tenfold would put most body cells out of diffusional reach of the nearest capillary, and redesigning the circulatory system to accommodate this counterfactual would pose intractable problems.

For example, one might try increasing the number of capillaries permeating the tissues so as to bring the cells back within "diffusional reach," but as Schmidt-Nielsen notes, the active muscles of a guinea pig already contain some 3,000 open capillaries per square millimeter of muscle (akin to 10,000 tiny parallel tubes running down the lead of a pencil).[15] This is a considerable number and occupies about 11 percent of the muscle's volume.[16] Even in large organisms such as a horse, with a lower metabolic rate than a guinea pig, muscles still contain 1,000 capillaries per square millimeter of tissue.[17] There is clearly a limit to the cross-sectional area of a tissue like a muscle which could be devoted to capillaries. Any greater density of capillaries and muscles would become watery sponges, mainly composed of blood. Increasing the diameter of individual capillaries would have the same effect. So, compensating for decreased rate of diffusion by significantly increasing the number of capillaries, or the diameter of individual capillaries, is not an option.

Another strategy to compensate for increased viscosity and decreased diffusion rates would be to maintain the same cross-sectional area of the capillary bed but shrink the individual capillaries to diminish the average distance between capillaries and tissue cells. However, this option also encounters a problem. According to the Hagen–Poiseuille equation, the flow of a fluid through a pipe varies in direct proportion to the radius of the pipe to the fourth power: $F = Kr^4$, where F is rate of flow and r is the pipe's radius. This means that halving the diameter of a capillary would necessitate increasing the pumping pressure sixteen fold to maintain the same flow.[18]

In effect a closed circulatory system of the sort used by large complex life forms on Earth would not be possible if water's viscosity were substantially higher and diffusion rates substantially less. The power required to pump blood would be prohibitively high, and compensatory redesigns to bring cells within diffusional reach would face intractable challenges.

Here again note that water's low viscosity is mainly of benefit to large complex macroscopic life forms for whom a circulatory system is essential. As I explained in the journal *BIO-Complexity*, "Some organisms, such as molds, can actually thrive at ambient temperatures in very viscous concentrated sugar solutions. It seems many types of unicellular life could thrive even if the viscosity of water were several times higher, but not complex metazoan organisms like ourselves."[19] The aptness of water's low viscosity for circulatory systems in complex macroscopic life forms (like us) is yet another indication that nature's fitness isn't just for the generic carbon-based cell but for beings of our physiological design—large complex multicellular organisms that require a circulatory system to supply their tissues with oxygen and nutrients.

If Water Were Less Viscous

Might water be fitter if its viscosity were lower and the diffusion rate higher? In such a counterfactual world, blood volume and the energetic costs of pumping would be reduced, but not significantly. However due to the r^4 constraint, even if the viscosity were distinctly lower, the capillary diameter and blood volume could only be cut by a relatively small fraction. Moreover, the cell's delicate cytoarchitecture would face more intense Brownian bombardment, since particle mobility in a fluid is inversely related to viscosity.[20] So, cellular structures would be far less stable. The half-lives of the cell's key macromolecules would also be decreased and the energetic burden of maintaining cellular homeostasis increased. It is highly doubtful if anything remotely resembling a living cell would be feasible if the viscosity of water approached, for instance, that of a gas.

Indeed, the low viscosity of gases is the reason biologists generally reject them as suitable media to instantiate a chemical living system.[21] Gases are far too labile to be serious candidates for a chemical matrix of life. Comments Arthur Needham:

Only systems based on a fluid medium could display the properties which we should accept as Life. Gaseous systems are too volatile and lack the powers of spontaneously segregating sub-systems... In gases... the tendency towards a uniform distribution of energy is very rapid but in liquids is slow enough for local differences to be maintained... and for steady-states to be set up. All gases mix freely with any other, but not all liquids. Some form discontinuities or interfaces, with solid-state properties, where they meet another liquid, and complex polyphasic systems are readily formed which further... increase the potentialities for steady-state perpetuation.[22]

Segregating sub-systems are unusual in a gas, but not unheard of. Clouds manage it, but as Needham notes, the very transience of cloud shapes points up the unsuitability of gas as a medium for stable segregating sub-systems.

There are additional adaptive constraints against a reduced viscosity. If blood were less viscous, as occurs in cases of anemia, blood flow would be more turbulent in the larger vessels in the vascular tree.[23] And greater turbulence has been implicated in predisposing to clot formation and damage to the endothelial lining of the arterial walls, eventually leading to atheroma.[24]

Surprisingly, increased turbulent flow would actually increase the pressure required to pump blood.[25] Paul Clements and Carl Gwinnutt explain:

Flow is less ordered and the eddy currents react with each other, increasing drag or resistance to flow. As a result, a greater energy input is required for a given flow rate when flow is turbulent compared to when flow is laminar [non turbulent]. This is best demonstrated by the fact that in turbulent flow, the flow rate is proportional to the square root of the pressure gradient, whereas in laminar flow [non-turbulent flow], flow rate is directly proportional to the pressure gradient. This means that to double the flow, the pressure across the tube must be quadrupled.[26]

Thus with a lower viscosity the cost of pumping would increase, at least in the larger vessels in the vascular tree.

Another disadvantage of the increased turbulence from lower viscosity is that, due to the more chaotic flow, the relationship between shear-stress level in arterial walls and flow rates would not hold, and the body's ability to maintain an optimal ratio of flow to vessel diameter by sensing changing stress levels in arterial walls and then remodeling the arteries accordingly (during athletic training, for instance) would cease to be feasible.[27]

It seems then that water's fitness for a circulatory system would likely be diminished if its viscosity were much less than it is. It is not possible to precisely specify the range of viscosities and diffusion rates compatible with a functioning circulatory system, but all the evidence suggests that the viscosity of blood must be very close to what it is. Viscosities of substances on Earth range over more than twenty-seven orders of magnitude, from the viscosity of air to the viscosity of crustal rocks.[28] Even setting aside these extremes, the viscosity of many common substances varies considerably.[29] The viscosity of water is 1; corn oil, 65; olive oil, 84; and glycerin, 1420.[30] Thus, the viscosity of blood, which is largely determined by that of water, sits in a tiny Goldilocks zone within the vast range of viscosities in nature.

Laplace's Law

The functioning of our vascular system also critically depends on an important physical law known as Laplace's law, a law which has nothing to do with the actual properties of water but which is nonetheless vital to the functioning of the circulatory system.

As noted above, a considerable pressure is needed to force the blood through the capillaries. The hydrostatic pressure at the arterial end of the capillary is 35 mm Hg, which is approximately a third of the pressure inside the aorta as the blood leaves the heart. One might have imagined that such a relatively high pressure applied to the capillaries would blow apart the capillary walls. After all, the capillary walls are formed from a fabric of ultrathin endothelial cells that wrap themselves around the lumen of the capillary, and the average width of these walls, which form the barrier between the blood in the lumen of the capillary and the tissues outside, is only a fraction of a micron.[31] (The capillary walls need to be very thin to minimize the barrier between the blood in the capillary and the tissues to enhance material exchange.) So how can such ultrathin walls withstand the considerable pressure of 35 mm Hg? The explanation is intriguing and is explained by Laplace's law.

Laplace's law states that the tension on the wall of a tube under pressure is $T = PR$, where T is the tension or stress on the wall of the tube, P is the pressure in the tube (really the difference between the pressure inside and outside the tube), and R is the radius of the tube.[32] This has surprising consequences. It means that the smaller the radius of a tube, the lower the tension in the wall of the tube and the more easily the tube can withstand internal pressure. So, bicycle tires (which have thin walls and a small radius) can be pumped to seven atmospheres while a car tire (thick walled with a large radius) can only be pumped to two atmospheres.

"The bottom line is that thanks to Laplace's law, thin pipes can have thin walls, certainly a convenience if exchange takes place in the smallest pipes of a transport system," Vogel comments.[33] And because the radius of the capillaries is very small, only between 5–10 microns, Laplace's law means that the tension on the walls is many times less than that on the walls of the aorta. If Laplace's counterintuitive law did not hold, the design of efficient closed circulatory systems like our own, which are predicated on narrow capillaries with thin walls, could never have been actualized, and we would not be here to contemplate the loss.

Physiological Determinism

CONSIDERATION OF the various design constraints that apply to the circulatory system—energetic, volumetric, diffusional, and so forth—lucidly reviewed in Vogel's *Comparative Biomechanics*[34]—leads to an intriguing conclusion: The only functional closed water-based circulatory system, the only design which will work for organisms of our physiological design, is one that conforms very closely to the design actualized in humans and in all large complex extant organisms on Earth. Given energetic and volumetric constraints, Murray's law[35] and Laplace's law,[36] the diffusion rates of oxygen and other key metabolites in aqueous solutions, and given the viscosity of water, one can predict capillary size, the density of the capillary bed, and even the design, in broad outline, of the whole cardiovascular manifold.[37]

In the words of researcher Cecil D. Murray, "No one can escape the impression of a physiological determinism as exemplified by the narrowness of the physiological range."[38] While the details vary among the various species that enjoy a closed circulatory system, there is only one main blueprint that will work, implying that the functional design of the cardiovascular system approaches that of a physical constant.

Given these considerations it is not surprising that capillary size does not, as Vogel notes, "vary in any systematic way with body size—capillaries of elephant and whale differ little from those of bat and shrew,"[39] and the volume of the blood and vascular tree take up the same fraction of the total volume of the organism in nearly all mammals.[40] Vogel continues: "Not only is the blood volume of an octopus, with its independently evolved circulatory system, similar to our own, but its capillaries and blood cells are also only a little larger than ours."[41]

As to the constancy of the fraction of body volume taken up by the cardiovascular system in various species, Schmidt-Nielsen comments, "The total volume of blood in most mammals is between 60 and 70 cm^3 blood per kilogram body mass; thus, relative blood volume is a size-independent parameter... that is, in general, the blood in mammals makes up a constant fraction of the body mass."[42]

The mass of the heart also occupies about the same fraction of the mass of the body in all mammals.[43] And further evidence of Murray's "physiological determinism" is that even the pressure required to pump the blood through the cardiovascular manifold in organisms as diverse as the giraffe and the mouse only varies by a factor of two. The mean arterial blood pressure of mice is 93 mm Hg; of humans, 100 mm Hg;

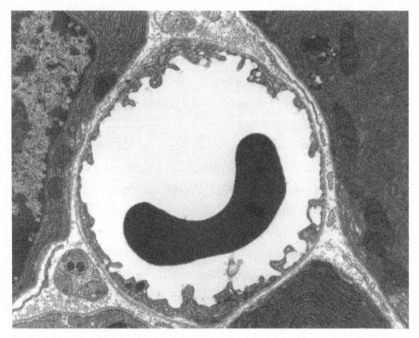

Figure 6.2. Red cell in mammalian pancreatic capillary.

of elephants, 150 mm Hg; and of giraffes, 200 mm Hg. And recall, the brain of the giraffe is up to two meters above its heart, creating an additional demand for increased blood pressure.[44]

Finally, another intriguing element of fitness: The necessary functional diameter of capillaries of about 5–10 microns is determined by various physical constants (e.g., diffusion rates in aqueous solutions, the viscosity of blood, the power of biological pumps). Additionally, the diameter of many cells is considerably larger than the diameter of the smallest capillaries,[45] and it is only through various adaptations to their shape and structure[46] that cells are able to squeeze through even the smallest capillaries. Imagine the difficulty cells would face in passing through the capillary bed if those physical constants had determined the necessary functional diameter of capillaries to be significantly less, say 1–3 microns in diameter. Circulation would grind to a halt.

The Right Volume

ONE VITAL characteristic of the canonical cardiovascular system, noted above, is that it occupies a relatively small and similar percentage of the total volume of the body in many organisms (as diverse as mammals and cephalopods). In an adult man this turns out to be about 5.4 liters,[47] or about 9 percent of body volume. If the functional volume in humans had to be two or three times greater to ensure the delivery of the 250 ml of oxygen to the tissues, then the body would be reduced to a spongy bag of fluid and the work of pumping rendered prohibitive, taking up 20–30 percent of the resting energy of the body. Moreover, there would be little room for "guts or gonads" (to borrow a phrase from Vogel), or for brains or any other organ systems, and the energetic burden would be insurmountable.

The fact that our cardiovascular system occupies a relatively small volume of the body is nothing short of a miracle given the number of very different physical parameters required to make such a modest functional volume possible. And this is the second miracle in this category. As we saw in the previous chapter, the functional volume of the lungs is about six liters in an adult man, also about 10 percent of the volume of the body,[48] sufficient to provide us with the 250 ml of oxygen we require each minute, thanks to an ensemble of pre-existing fitness parameters in tandem with certain ingenious adaptations. As we saw, if it required twice the volume to deliver the same quantity of oxygen, fitting the lungs in the body would have necessitated a radical redesign of the human body plan.

What is true of our lungs is equally true of our circulatory system. If the functioning of these two organ systems had necessitated only twice the volumes to serve their respective functions, then our upright android design, as well as the design of all other large terrestrial aerobic organisms from elephants to shrews, would have been massively constrained.

No Conceivable Alternative

As IN so many other cases of prior environmental fitness for our biological being (for life on the land, for oxygen generation via photosynthesis, for a safe oxygen-enriched atmosphere, for air-breathing, et cetera), water's supreme fitness to form the medium of the blood and enable our circulatory system depends not just on one element of natural fitness but on an ensemble of elements working together. There is water's generic property of incompressibility, its peerless powers as a solvent, its low viscosity, the relatively high diffusion rate of solutes in aqueous solutions, its low density, and so forth.

That water is so profoundly and uniquely fit to form the medium of the circulatory system is highly fortunate, for there is no alternative fluid which might stand in for water to perform this vital role in complex multicellular carbon-based organisms. Since the advent of complex multicellularity 500 million years ago, no organism on Earth has found any alternative to water upon which to base its circulatory system.

Moreover, from the evidence reviewed in this chapter, it is clear that there is only one possible functional design for a cardiovascular system based on water, and this is largely determined by the properties of the wonder fluid itself. If the various physical constants involved in determining the function of the cardiovascular system (e.g., the viscosity of water, the density of water, diffusion rates in aqueous solutions, Laplace's law, Murray's law) were even modestly different from what they are, no functional cardiovascular system based on water could ever have been actualized. And as water is the only candidate available, without the environmental ensemble of fitness manifest in the properties of water, no large multicellular carbon-based organisms would exist, and we would not be here to contemplate the loss.

The fitness of this wonder fluid for the design of a functional circulatory system is like the fitness of the atmosphere for safe air-breathing. It's a fitness *for us*, for our type of complex multicellular carbon-based life, a fitness irrelevant to the hordes of unicellular and simple multicellular organisms which comprise the vast majority of life forms on Earth.

7. WARM-BLOODED

One does not like to accept a fact of such far-reaching importance as mere chance, and yet no other explanation was at hand. For, after the briefest consideration, it was obvious that here, at least, natural selection could not be involved. But it was also certain that this is no unique instance of a property of a simple substance automatically serving a very useful purpose in the processes of life. Like every one who has received a chemical training, I was vaguely conscious of numerous other similar cases; like every one who has any acquaintance with the general properties of matter, I knew that the remarkable thermal properties of water are of great importance to living organisms. However, in spite of the fact that I had been brought face to face with a definite problem whose solution now appears to be perfectly patent, so great is the natural inertia of the mind, and so firmly established was the belief that natural selection is, on the whole, quite adequate to account for biological fitness, that for a number of years I made no further progress.

—Lawrence Henderson, *The Fitness of the Environment* (1913)[1]

ONE OF THE SEVERAL CHALLENGES THAT ALL TERRESTRIAL ANI-mals face is the much greater variability in temperature on land compared to the sea. Many terrestrial environments (unlike aquatic) undergo considerable diurnal temperature changes, and in the higher latitudes annual changes in temperature may be as great as 100°C. As the author of an early text on warm-bloodedness, William Hale-White (physician to Guy's Hospital in London in the late nineteenth century) pointed out:

> The temperature of the air at the same spot may vary as much as 30° Fahrenheit in a few hours, and the variation for the year is often 100° or more. The temperature of water is much more constant... While terrestrial animals are subjected to extreme and often sudden changes of temperature, aquatic animals, on the other hand, are never subjected to such extremes, for the annual changes in the temperature of their surrounding medium are slight; they are also very gradual, and may to a certain extent be avoided by alteration of locality... The change [during terrestrialization] must have been a very severe one, for the various vital and chemical processes, which had hitherto been carried on in a medium whose temperature was not liable to sudden alterations, and whose variations in temperature were slight.[2]

Two different strategies are available to deal with terrestrial temperature fluctuations. The first and by far the simplest is to "go with the flow" and allow body temperature to vary with the environmental temperature—a physiological trait known as ectothermy—the physiological strategy of amphibia, reptiles, and most terrestrial invertebrates.[3] Ectotherms are often referred to as cold-blooded.

An alternate strategy, endothermy or homeothermy, buffers the body against temperature swings by regulating body temperature so as to maintain it at a fixed level. In the case of extant mammals and birds, this fixed level is between 37°C and 41°C, considerably above Earth's average surface temperature of about 15°C and much higher than the environmental temperature in the higher latitudes.[4] Thus we, along with our mammalian cousins and birds, are commonly referred to as warm-blooded.

Endothermy has many biological advantages. The most obvious is that organisms that maintain a constant body temperature can be active day and night, winter and summer. Some warm-blooded animals maintain a body temperature nearly 100°C above the environment, such

Figure 7.1. A polar bear traverses the ice in the Alaska portion of the southern Beaufort Sea.

as penguins in the Antarctic and male polar bears overwintering on the Arctic sea ice.

As Michael Angilletta notes, warm-blooded organisms appear to benefit from precise thermoregulation in other ways as well.[5] For example, maintaining a particular near-constant temperature allows a range of enzymatic and catalytic activities and other processes in the body to function optimally.[6]

As a general rule, the higher the normal body temperature of a species, the higher its rates of enzymic activity, energy generation, growth rate of cells, rate of development, rate of movement in animals, and rate of photosynthesis in plants. Reproductive fitness across a broad range of organisms is also increased.[7]

Moreover, many cold-blooded animals do not achieve anywhere close to the maximal performance their muscles are capable of delivering when active at low body temperatures. And this is why many cold-blooded organisms, including flying insects such as bumble bees and dragon-

Figure 7.2. The Permian reptile Dimetrodon, with an Edaphosaurus in the background.

flies, are warm-blooded during flight and must warm up by shivering before taking off in colder climates.[8]

Many warm-blooded animals also show a reduction in muscle performance when operating in lower external temperatures. Studies of muscle performance reveal the advantages of being warm. These show, for example, that the measured rates of force development of muscles increase with external temperature.[9] Studies also show that nerve conduction rates in the human hand and arm increase up to two meters per second per degree as the skin temperature rises from 5°C to 37°C.[10]

Cold-Blooded Adaptations for Warming Up

The advantages of being warm-blooded is also observed in the efforts by ectotherms to increase their body temperatures. Crocodiles have blood-filled scales on their back, which act as solar panels to warm the animals on cold days.[11] The Permian reptile Dimetrodon used a large sail on its back, which acted as a heat absorber to raise its body temperature several degrees. The sail could be pointed sunward for rapid warming, perhaps allowing Dimetrodon to hunt before its prey was active. C. D. Bramwell and P. B. Fellgett calculate that without the sail it would have taken a 250 kg (440 lb) Dimetrodon more than three hours to raise its body temperature from 26°C to 32°C (from 79°F to 90°F).[12]

Another sign of the advantage of warm-bloodedness: some predatory ocean fish maintain their brain, nervous system, and eyes at temperatures several degrees higher than the temperature of the rest of their bodies, which track more closely the temperature of the surrounding sea water.[13] The authors of a paper in *Current Biology* elaborate:

> Large and powerful ocean predators such as swordfishes, some tunas, and several shark species are unique among fishes in that they are capable of maintaining elevated body temperatures (endothermy) when hunting for prey in deep and cold water. In these animals, warming the central nervous system and the eyes is the one common feature of this energetically costly adaptation. In the swordfish (Xiphias gladius), a highly specialized heating system located in an extraocular muscle specifically warms the eyes and brain... above ambient water temperatures.... Warming the retina significantly improves temporal resolution, and hence the detection of rapid motion... temporal resolution can be more than ten times greater in these fishes than in fishes with eyes at the same temperature as the surrounding water.[14]

It is true that some cold-blooded organisms exhibit impressive abilities. For example, some fish use tools; one species has been observed communicating effectively to another aquatic species via gestures in a bid for cooperation; and the octopus is capable of quite complex problem solving.[15] Yet given the evidence that neural processing is enhanced at higher temperatures,[16] it seems unlikely that the intelligence of cold-blooded species can equal that of warm-blooded organisms. All the most highly intelligent organisms in the seas and on the land, such as orca, corvids, parrots, and primates, are air-breathing endotherms that maintain body temperatures many degrees above the surrounding environment. Moreover, the fact that many different lineages of birds and mammals have achieved endothermy and maintained it over millions of years of evolution despite the much greater metabolic cost suggests that it has very significant physiological advantages.

We have seen that the need for oxidations is a universal requirement for complex intelligent carbon-based life and will surely apply throughout the cosmos.[17] We also have seen that the highest metabolic rates can only be achieved by terrestrial organisms obtaining their oxygen directly from an atmosphere and that this likely will also apply throughout the cosmos. The likelihood that endothermy will be a universal requirement for high intelligence is also a reasonable conjecture.

Water's Thermal Properties

WE SAW in Chapter 2 that it was an ensemble of prior properties of water more than anything else that, by enabling the hydrological cycle and the chemical weathering and erosion of the rocks, created the right terrestrial conditions for life to colonize the land. And we saw in subsequent chapters that another remarkable ensemble of fitness enables air breathing by advanced metabolically active terrestrial organisms like ourselves. But there is yet another ensemble of environmental fitness, this one involving the thermal properties of water, that is essential for endothermy in terrestrial organisms like ourselves. These thermal properties are water's high specific heat capacity, its high latent heat of evaporation, and its high heat conductivity.

Specific Heat

Water's specific heat capacity is higher than any familiar liquid except liquid ammonia.[18] This means that a body of water changes temperature less with the addition or subtraction of a given quantity of heat than do most other substances, contributing to what Tom Garrison calls water's "thermal inertia."[19] One need only take a dip in a lake or sea on a cold night after a warm summer day to experience the temperature stability of water. At noon on a hot day, the body of water at, say, 24°C (75°F) feels cool; but come night, as the air grows chilly, the water's temperature has held more or less steady at the temperature it was at noon.

Water's uniquely high thermal inertia is an obvious vital element of fitness for all endotherms, making it easier for them to maintain a constant body temperature. The vital benefit of water's high thermal inertia was made clear some time ago by Lawrence Henderson:

> The living organism itself is directly favored by this same property of
> its principle constituent [water], because a given quantity of heat pro-

duces as little change as possible in the temperature of its body. Man is an excellent case in point. An adult weighing 75 kilograms... when at rest produces daily about 2400 great calories which is an amount sufficient to raise the temperature of his body by more than 32° centigrade. But if the heat capacity of his body corresponded to that of most substances the same quantity of heat would be sufficient to raise his temperature between 100° and 150°. In these conditions the elimination of heat would become a matter of far greater difficulty, and the accurate regulation of the temperature of the interior portion of his body during periods of great muscular activity, well-nigh impossible. Extreme constancy of body temperature is of course a matter of vital importance at least for all highly organized [endothermic] beings, and it is hardly conceivable that it should be otherwise.[20]

As Henderson notes, our metabolism, which is set for maximum efficiency at a particular temperature, would be greatly stressed by large temperature swings, leading as they would to significant differences in the rates of enzymatic and other types of chemical reactions. Thus, for example, a ten-degree centigrade increase in temperature "will more than double the rate of a chemical change."[21]

The vital importance of temperature constancy in endotherms is highlighted by the fact that in humans any increase in body temperature much above 37°C (98.6°F) is dangerous. An increase in body temperature much above 40°C (104°F) can be fatal. However, thanks to water's high thermal inertia, cases of death by hyperthermia are exceptional, even in hot climates.

The high specific heat of water also benefits terrestrial ectothermal organisms, because although their body temperatures "go with the flow," the rapidity of their bodily temperature changes is greatly attenuated by the high specific heat of water.

Would a higher heat capacity be even more beneficial to life? There are reasons to doubt this. For one, ectotherms might be severely disadvantaged if the heat capacity of water were higher. We have all seen bumblebees and dragonflies shivering to warm their thoracic flight muscles before flying on a cold morning.[22] If water's specific heat were any higher, ectotherms would have an even harder time warming themselves on cold days.

What about for warm-blooded organisms? An intriguing hint that if the specific heat of water were higher its fitness might be lower is that the body temperature of warm-blooded organisms is close to the temperature where the specific heat of water is least, about 36°C.[23] Arthur Needham comments on this:

> It is probably highly significant that the body temperature of homeothermic [warm-blooded] vertebrates is around 37.5°C, and that large insects warm up to this temperature before taking flight. This is precisely the temperature at which the specific heat of water is minimal, so that its molecules are then most easily mobilized per unit energy supplied. Compost and damp hay tend to this temperature, under bacterial action, though in large artificial masses the reactions may then get out of control and the temperature may soar to firing point.[24]

Lastly, it is worth also noting that the high specific heat of water works to buffer organisms against rapid changes in body temperature because about 60 percent of the mass of our bodies is water and most of the tissues have specific heat not much less than that of water.[25] This is an important additional element of natural fitness for endothermy. That the matrix of life makes up 60 percent of the body, and most of the tissues have specific heats not much lower than water, are major elements of prior natural fitness enabling both endothermic and ectothermic terrestrial life.

Latent Heat of Evaporation

It is remarkable enough that one of the thermal properties of water is so ideally fit for terrestrial endothermy. It is doubly remarkable that another unique thermal property of water—its very high latent heat of evaporation—is also ideally fit for terrestrial endothermy.

The latent heat of evaporation of water, one of the highest of any liquid, has a dramatic cooling effect. As Schmidt–Nielsen explains, it takes one hundred calories to heat a gram of water from freezing to boiling, which is no small amount. And yet it takes more than five times this amount, some 584 calories, to transfer a gram of room-temperature water to water vapor at the same temperature.[26]

More than a century ago Henderson considered water to have the highest known latent heat of evaporation of any fluid,[27] and fifty years later Needham claimed the same.[28] We now know of several substances with a higher latent heat of evaporation, but these are mainly liquid metals at very high temperatures,[29] and water's latent heat of evaporation remains "greater than any other known molecular liquid."[30]

Evaporative cooling plays an essential role in ridding the body of excess heat in hot weather and during strenuous exercise, particularly for endotherms like birds and mammals, whose very high metabolic rate generates great quantities of heat. Henderson elaborates:

> In an animal like man, whose metabolism is very intense, heat is a most prominent excretory product, which has constantly to be eliminated in great amounts, and to this end only three important means are available: conduction, radiation, and evaporation of water. The relative usefulness of these three methods varies with the temperature of the environment. At a low temperature there is little evaporation of water, but at body temperature or above there can be no loss of heat at all by conduction and radiation and the whole burden is therefore thrown upon evaporation.[31]

Indeed, at environmental temperatures close to and above normal body temperature, conductive and radiative heat loss from the skin reverse as the body starts to soak up heat from the environment.[32] In such cases, as is frequently the situation in hotter climates around the globe, maintaining a constant body temperature of 37°C (98.6°F) would prove impossible without the dramatic cooling effect of the evaporation of water. And even in temperate regions of the earth, the heat generated in strenuous exercise would quickly risk fatal hyperthermia. Warm-bloodedness would be impossible without the dramatic cooling effect of the evaporation of water, except perhaps for sedentary organisms living permanently in cool climates.

An important point that Henderson and many subsequent commentators generally pass over is that evaporative cooling only directly benefits terrestrial organisms. The evaporative cooling effect is irrelevant to aquatic organisms, whether large or small, whether air-breathing or water-breathing. Consequently, the high latent heat of evaporation is an element of environmental fitness specifically for terrestrial organisms. And because of our copious sweat glands and lack of significant body hair over much of our bodies, water's unusually high latent heat of evaporation is arguably of maximal utility to modern humans compared with any other extant organism on Earth. This is why in hot climates human endurance hunting is a successful hunting strategy.[33]

Although human bipedal locomotion is more energy efficient than running on all fours,[34] and humans have the ability to carry a water supply with them on the chase, more than any other factor it is the nakedness of the human body along with our copious sweat glands which together maximize the evaporative cooling effect of water on the skin. This greatly increases heat loss above what can be achieved by animals covered in thick furry hides with few sweat glands and reliant on panting. Although a hunting dog can outpace a human over 15–20 minutes, in an endurance race of say ten kilometers in the heat of the day, fit humans, thanks to our ability to maximally exploit the evaporative cooling of water, can leave hunting dogs literally "for dead."[35]

A telling glimpse of a counterfactual world without evaporative cooling is provided by the disease anhidrosis, involving dysfunctional sweat glands, which diminishes or destroys the body's ability to shed heat by sweating and which commonly leads to heat exhaustion following strenuous exercise and, in the worst cases, death.[36]

In short, in hot climes humans absolutely depend on water's outsized powers of evaporative cooling. If a magic wand were waved tomorrow and water's high latent heat of evaporation vanished, billions of individuals in the tropics would die within a few hours.

Another important point to note in this context is that the cooling effect of water evaporation can only occur when the air is not saturated with water vapor—that is, when the humidity is less than 100 percent. A recent paper in the journal *Science* explains:

Humans' bipedal locomotion, naked skin, and sweat glands are constituents of a sophisticated cooling system. Despite these thermoregulatory adaptations, extreme heat remains one of the most dangerous natural hazards, with tens of thousands of fatalities in the deadliest events so far this century....

While some heat-humidity impacts can be avoided through acclimation and behavioral adaptation, there exists an upper limit for survivability under sustained exposure, even with idealized conditions of perfect health, total inactivity, full shade, absence of clothing, and unlimited drinking water. A normal internal human body temperature of 36.8° ± 0.5°C requires skin temperatures of around 35°C to maintain a gradient directing heat outward from the core. Once the air... temperature... rises above this threshold, metabolic heat can only be shed via sweat-based latent cooling, and at... [100 percent humidity and temperatures] exceeding about 35°C, this cooling mechanism loses its effectiveness altogether.[37]

Currently on Earth, few places approach the fatal combination of temperatures of 35°C and 100 percent humidity, and these are coastal regions in subtropical areas where the surface temperature of the sea may exceed 35°C. As the authors comment, "All are situated in the subtropics, along coastlines (typically of a semi-enclosed gulf or bay of shallow depth, limiting ocean circulation and promoting high... [sea surface temperatures], and in proximity to sources of continental heat, which together with the maritime air comprise the necessary combination."

But the fact that such places comprise only a tiny fraction of Earth's surface is a close call. If the configuration of oceans and continental land masses had been different, or if the circulation of the oceanic waters had been less, our sun a little brighter, our planet a little closer to its host star, or our atmosphere appreciably more given to retaining the heat of the sun, then regions that experience such high temperatures and high humidity might have been much more common and, despite the cooling effect of evaporation, the earth a far less clement habitat for humans.

Thermal Conductivity

The third thermal property of water enabling endothermy is its high heat conductivity compared with many other liquids. This, as Henderson pointed out, favors "the equalization of temperature within living cells whose structure hinders the establishment of convection currents."[38]

The conductivities of different materials, as is the case of so many physical parameters, range over several orders of magnitude. Air has a conductivity 40,000 times less than silver.[39] Some poor conductors— e.g., dry air and wood—are more properly termed insulators. Fluids generally have a very low conductivity of heat compared with many solids, although among familiar liquids water has the highest.[40] Only liquid metals have higher thermal conductivities.[41]

As mentioned above, warm-blooded creatures like ourselves produce great quantities of metabolic heat, which must be rid from the body to prevent fatal hyperthermia. The heat, produced in the mitochondria where the oxidation of the body's fuels occurs, is conducted from the tissue cells into the blood capillaries, a process greatly aided by water's high thermal conductivity. Thermal conductivity, being a form of diffusion, is very rapid over short distances, which is exactly what is needed to rapidly carry heat the short distance from the tissues to the blood in the capillaries, thereby preventing a fatal buildup of heat in actively metabolizing tissue cells. This rapid rate of conductivity over short distances is also exactly what is needed to transfer heat from the capillaries in the periphery the short distance to the surface of the skin, where water's high latent heat of evaporation draws heat out of the body.

If water's conductive powers were much less—on par with substances such as cotton, wool, or wood—heat could not be transferred quickly enough from the tissues to the blood, especially in situations of strenuous exercise. The tissues would boil.[42] On the other hand, if the conductivity of heat by water were much greater, say like that of silver or copper, environmental temperature swings would be rapidly transferred (as along a metal rod) throughout the body and swamp the thermal inertia afforded by water's high heat capacity. We would suffer deadly swings in body temperature, rendering terrestrial endothermy untenable.

For terrestrial endotherms (and to a degree ectotherms also) the heat conductivity of water must lie somewhere close to what it is to ensure the required level of thermal stability. While a higher heat conductivity might benefit us when we were "on the run" (by accelerating heat excretion), on cool nights we would be prone to lose heat too rapidly and in colder climes even freeze to death.

Water's heat conductivity is fit in another intriguing way for thermal stability. "For most liquids, the thermal conductivity (the rate at which energy is transferred down a temperature gradient) falls with increasing temperature," Martin Chaplin explains, but in the case of water, the opposite holds until temperatures reach well above the ambient temperature range. In the ambient temperature range, "if our cells produce excess energy, that heat energy is transported away more efficiently at higher temperatures, reducing its heating effect," he continues. "At lower temperatures, with lower thermal conductivity, the heat is less well transported away, allowing greater heating effect. Thus our cells are more able to stabilize their temperature."[43]

One consequence of water's high heat conductivity is that it readily conducts heat away from the surface of aquatic organisms and renders warm-bloodedness in the sea energetically costly. This is another reason that despite the great advantages of being warm-blooded—faster muscles, faster swimming speeds, faster neurological processing and response times, much faster visual temporal resolution—there is only one water-breathing organism known to have achieved whole-body endothermy in 500 million years of evolution, the opah fish mentioned in Chapter 4. But in the case of the opah, body temperature is maintained only a few degrees C above the surrounding water. This is a very modest degree of endothermy compared with that of mammals and birds, where many species maintain their body temperature at 50–100°C above their surroundings.

The level of endothermy achieved in mammals and birds would appear to be wholly beyond the reach of aquatic water-breathing creatures. First, as discussed above, the heat generated in an aquatic organism is readily drawn away into the surrounding medium due to the high heat conductivity of water. Second, because water-breathing is a much less efficient way to obtain oxygen, the metabolic rates and consequent rates of heat generation in water-breathing organisms is far below that attained by air-breathing sea-goers such as whales, penguins, and seals. Moreover, all warm-blooded air-breathing marine species have thick protective layers of insulation—blubber, feathers, fur—which greatly insulates them against conductive heat loss in water.

Finally, a fourth element of fitness for terrestrial endothermy is the low heat conductivity of air—twenty-five times less than water's.[44] This is another crucial element of prior environmental fitness for terrestrial endothermy. If the conductivity of air were similar to that of water, then just as the high heat conductivity of water mitigates against aquatic endothermy, the high conductivity of air would make terrestrial endothermy impossible. Heat would be continually drawn from the body, and maintaining a constant warm temperature would impose an impossible metabolic burden on the organism.

Water: The Gift of Terrestrial Endothermy

THE SECRET of terrestrial endothermy resides in what is surely on any consideration a remarkable ensemble of environmental fitness: the thermal properties of water. Its specific heat, latent heat of evaporation, and heat conductivity work together to enable terrestrial endothermy. Had their values not been very close to what they are, warm-blooded terrestrial organisms would never have emerged. No amount of Darwinian trial and error could have created terrestrial endothermy without the prior environmental fitness of the thermal properties of water.

That endothermy in terrestrial organisms should be enabled by all the thermal properties of water working in synergy provides further compelling evidence that the laws of nature are uniquely fit for advanced terrestrial carbon-based life forms like ourselves. If you wanted to play Plato's demiurge and, starting from scratch, create an ensemble of fitness enabling terrestrial warm-blooded carbon-based life forms, beings of our physiological design, you would need to create water and configure it with precisely its present suite of thermal properties.

8. Oxygen: Delving Deeper

Why grass is green, or why our blood is red,
Are mysteries which none have reach'd unto.
> —John Donne, "An Anatomy of the World" (1611)[1]

Chlorophyll, for example, contains magnesium, and it is thought that the process of reduction in the leaf may depend upon the characteristic of this element.... In like manner, haemoglobin contains iron, and the capacity of haemoglobin to unite with oxygen and as oxyhaemoglobin to carry it from the lungs to the tissues is unquestionably due to the chemical behavior of that metal.
> —Lawrence Henderson, *The Fitness of the Environment* (1913)[2]

Transition metal inorganic elements in some organization are of the essence of life as much as this is true of amino acids and nucleotides.
> —Robert J. P. Williams, "The Symbiosis of Metal and Protein Function," (1985)[3]

THE PREVIOUS CHAPTERS PROVIDED CLEAR EVIDENCE FOR A SPE-cial fitness in nature for advanced terrestrial air breathers like ourselves. In this chapter we will examine several additional elements of fitness which enable aerobes of our physiological design to utilize oxygen.

First, recall that one crucial element of fitness is the attenuation of oxygen's reactivity at ambient temperatures, which occurs due to a special feature of the oxygen atoms' outer electron shells, technically called "spin restriction."[4] This plays a critical role in saving us from spontaneous combustion. This key element of fitness not only enables us to live safely in an atmosphere greatly enriched in oxygen but, because air breathing is so much more efficient than water breathing for acquiring oxygen, it also enables air breathers to achieve much higher metabolic rates than water breathers such as fish and aquatic invertebrates. And this in turn enabled terrestrial air breathers to attain endothermy as well as high intelligence and sophisticated behaviors.

But this comes at a price: it means that activating and using atmospheric oxygen (O_2) demands special chemical mechanisms to overcome its unique reluctance to react. All the elements of fitness in nature for oxidative metabolism would be of no avail unless oxygen's reluctance to react at ambient temperatures could be overcome and its huge potential energies used to empower our type of advanced aerobic life. Happily, nature has obliged by providing a set of atoms—the transition metal elements—with just the right electronic structure and properties to activate oxygen.

The fact that all oxygen-handling molecules in nature employ the transition metal atoms, and no alternative atoms or organic molecules have been found to replace them in billions of years of evolution, suggests that the electronic properties of the transition metals are uniquely fit for this role and that without them aerobic biology would not exist, neither here nor anywhere else in the cosmos.[5]

One Electron at a Time

In his wonderfully lucid book *Oxygen: The Molecule that Made the World*, Nick Lane notes that there are only two means of overcoming oxygen's reluctance to react. One, oxygen can absorb energy from neighboring molecules excited by light or heat, such as in a campfire. Or two, it can be fed electrons one by one so that each unpaired electron in a given oxygen atom receives a partner independently. The transitional metal atoms like iron and copper have the ability to do precisely this. Iron, for example, can donate one electron at a time and commence the activation of oxygen because it possesses its own unpaired electrons. "Iron loses these electrons without becoming unstable because it has several differ-ent 'oxidation states,' all of which are stable under relatively normal con-ditions," Lane writes. This is partly due to the fact the iron atom is large and the electrons farthest from the nucleus are bound loosely to the iron nucleus. "The ability of iron to feed electrons one at a time," Lane says, "explains its affinity for oxygen." It also explains the body's need to "lock away iron in molecular safe houses," such as hemoglobin, to prevent the formation of active oxygen species. "Other metals that exist in two or more stable oxidation states, such as copper, can feed electrons just as efficiently," Lane adds, "and are equally dangerous unless well caged."[6]

In fact, all the oxygen-activating enzymes make use of transition metal atoms.[7] In a campfire the essential inertness of oxygen at ambient temperatures is rescued by the application of heat to raise the tempera-ture sufficiently to overcome the energy barrier to reaction and ignite the fire. In the body, oxygen's inertness is rescued by the unique ability of the transition metal atoms, like iron, to donate one electron at a time to the oxygen molecule (O_2), commencing its activation and reduction for en-ergy generation in aerobic organisms like ourselves (and all other aerobic organisms—whose high metabolic rates and "active lifestyles" depend critically on the energy of oxidations).[8] If not for this unique ability of the transition metal atoms to donate one electron at a time to an oxygen molecule, the tremendous energies of oxidation would be unavailable to carbon-based life forms, and neither humans nor any advanced life forms would exist on Earth, or indeed, anywhere in the universe.

Some intriguing work carried out by a research group led by Ulrike Diebold at the Institute for Applied Physics at the Vienna University of Technology using an atomic force microscope has enabled researchers to mimic the ability of transition metal atoms to add a single electron to an oxygen molecule. As team member Martin Setvin explains (echo-ing Nick Lane), in order to render oxygen chemically active, "you can increase the temperature which happens when you burn things," or "you can add an additional electron to the molecules."[9] Because organisms cannot use fire to activate oxygen, the lone alternative of activating each oxygen molecule by first adding one electron is ubiquitous in biological systems. As Setvin explains, "All living organisms use this trick."[10] And now using the atomic force microscope, it's possible to carry out the same trick in the lab. The Phys.org article quoting Setvin and the professor heading up the research group, Ulrike Diebold, explains:

> Setvin and coworkers are now able to activate individual O_2 molecules at will using a force microscope, and learn how the process occurs at the atomic scale. In the experiments, oxygen molecules were studied on the surface of a titanium oxide crystal at extremely low temperatures... "A tiny needle is vibrated and moved across the surface. When the atoms at the very end of the tip come close to the surface, the tip feels a force and the oscillation changes. From this tiny change, one can create an image showing where the atoms are," says Diebold. "Essentially, the reactive oxygen molecules that have an extra electron exert a stronger force on the tip than the unreactive ones, and thus we can distinguish them." Interestingly, it is also possible to inject an additional electron to an individual oxygen molecule with the same tip, and then observe the transition from the inactive to the active state.[11]

And note again, assuming as we must that carbon-based life is almost certainly the only form of chemical life ordained by nature, and assuming, as also seems likely, that all complex advanced carbon-based life will employ oxidations to derive energy,[12] then it follows that throughout the universe all intelligent species will not just be aerobic and terrestrial air-breathing life forms, but also will use the unique abilities of the transition metal atoms to activate oxygen.

In sum, there is hot activation of oxygen by fire and "cold" activation in the body by iron (or another transition metal). It is the latter method that enabled us to attain a high metabolic rate and, with it, a sufficiently high intelligence to master oxygen's hot activation, which launched our ancestors on the long road via fire and metallurgy to our modern technological civilization.

The role of the transition metals in biology in both the activation of oxygen and in conducting electrons along electron transport chains was dealt with at length in one of my previous books, *The Miracle of the*

Cell. An interested reader may browse through that monograph or look at relevant chapters in some of the main texts on bioinorganic chemistry.[13] Despite the somewhat technical nature of this area, I have adapted the short section on cytochrome C oxidase below from *The Miracle of the Cell*, as this enzyme illustrates perhaps more forcibly than any other oxygen-handling enzyme just how central the transition metals are in oxidative metabolism.

Cytochrome c Oxidase

CYTOCHROME C oxidase is the terminal member of the electron transport chain (ECT) in the mitochondria and performs the critical final reaction of oxidative metabolism, which involves transferring the electrons flowing down the ETC in the mitochondria to molecules of dioxygen (O_2), reducing them to water. The Cytochrome c oxidase enzyme sits astride the inner bilayer lipid membrane in the mitochondria, an enzyme that Earl Frieden identifies as probably the single most vital enzyme to all aerobic life:

> This is the enzyme, found in all aerobic cells, which introduces oxygen into the oxidative machinery, that produces the energy we need for physical activity and chemical synthesis... This enzyme may be regarded as the ultimate, in the integration of the function of iron with copper in biological systems. Here in a single molecule, we combine the talents of iron and copper ions to bind oxygen, reduce it with electrons from the other cytochromes in the electron transport chain, and finally to convert the reduced oxygen to water.[14]

The electrons flow within the molecule along a "transition metal wire" composed of a succession of iron and copper atoms that conduct them to the final catalytic center where they reduce oxygen to form water.[15] Here we see, in one vital protein complex, the unique ability of transition metals to draw electrons along an energy gradient and adroitly donate electrons one at a time to reduce oxygen via a set of partially reduced intermediates to water.[16] And among the myriad of different organisms that use this key enzyme, none has found other substances to replace the roles of the transition metals iron and copper in the billions of years since the onset of oxidative metabolism.

Cytochrome c oxidase also contains two other metal atoms—zinc and magnesium. They are not directly involved in transferring electrons or reducing oxygen to water. Zinc appears to play a structural role away from the active site, while magnesium may be involved in releasing the water molecules at the active site.[17]

Thus the work of this remarkable nano-machine, whose basic structure is built up of carbon, nitrogen, oxygen, sulfur, and hydrogen, depends on the unique properties of four metal atoms: iron and copper, involved in its core function, and zinc and magnesium, in supporting roles.[18]

The Fitness of the End Products

WITH THE reduction of oxygen to water by cytochrome c oxidase (O_2 + 4e +4H = $2H_2O$) in the mitochondria, we come to the end of the oxygen cascade that starts with oxygen entering the respiratory tract. But this is by no means the end of the ensemble of unique fitness in nature for oxidative metabolism. To understand the additional ensemble of fitness after the final reduction of oxygen to water, recall the basic chemical reaction which sustains us air-breathing aerobes:

Reduced carbon (CH)+oxygen → CO_2+H_2O+heat+energy (ATP).

The mutual fitness of the three waste products of this reaction (carbon dioxide, water, and heat) to work together to excrete two of the waste products, carbon dioxide and heat, is so remarkable that on any serious consideration it should finally banish any doubts that a skeptic might harbor as to the unique fine tuning of nature for our type of life—air-breathing terrestrial aerobes.

Carbon Dioxide

Consider first CO_2. To begin with, this waste product has a whole suite of properties which enable its excretion. In an article for the journal *BIO-Complexity*, I made the point that had CO_2 been a toxic substance, or a liquid insoluble in water, or an insoluble solid, or dissolved in water to

100 μm 150x

Figure 8.1. Grains of sand under an electron microscope.

form a strong acid, the reaction could never have been completed and the complete oxidation of carbon to carbon dioxide could never have been exploited by us energy-hungry aerobes.[19] However, CO_2 is none of these things. On the contrary, it is a relatively innocuous unreactive compound and very significantly a gas at ambient temperatures.

That it is a gas is highly fortuitous, since it is one of the very few gaseous oxides at ordinary temperatures.[20] Chemist R. T. Sanderson emphasizes this point:

> In a pensive mood the fireman who keeps the furnace stacked with coal might feel grateful that carbon dioxide rises through the stack and does not need to be shoveled out with the ashes. Otherwise the question of why carbon dioxide is a gas rather than a polymeric solid comes to mind. When we learn that silicon resembles carbon it may arouse some curiosity as to why carbon dioxide is so different from sand [silicon oxide]... If CO_2 was a polymeric solid, each carbon atom would be surrounded by four oxygen atoms and each oxygen also to another carbon

atom. In such a polymer, the bonding would have to be by single bonds. It is clear from the non-existence of this polymer that one double bond is more stable than its equivalent in two single bonds in CO_2.... And this is the reason why carbon dioxide does not polymerize [into a solid like silica].... If carbon single bonds had been much stronger, CO_2 would have been like sand![21]

As mentioned in Chapter 5, CO_2's gaseous nature at ambient temperatures is a property crucial to air-breathing organisms, since it means it can be readily excreted in the lungs of terrestrial organisms via the respiratory tract—the same route through which oxygen is absorbed. Henderson elaborates:

> In the course of a day a man of average size produces, as a result of his active metabolism, nearly two pounds of carbon dioxide. All this must be rapidly removed from the body. It is difficult to imagine by what elaborate chemical and physical devices the body could rid itself of such enormous quantities of material were it not for the fact that... in the lungs... [carbon dioxide] can escape into air which is charged with but little of the gas. Were carbon dioxide not gaseous, its excretion would be the greatest of physiological tasks; were it not freely soluble, a host of the most universal physiological processes would be impossible.[22]

But the great convenience to aerobes of CO_2's gaseous nature in ambient conditions is only the first of many elements of fitness exhibited by this remarkable substance. Another is the way CO_2 is transported to the lungs. As every medical student learns, despite the fact that the solubility of CO_2 is many times that of oxygen, most of the CO_2 is transported, not in simple solution, but as bicarbonate. Although its excretion is as a gas, its transport in the blood from tissues to lungs is mainly as a nongaseous soluble compound, bicarbonate.

Bicarbonate is formed in the blood when the CO_2 produced from oxidation of the body's reduced carbon fuels reacts with water:

$$H_2O + CO_2 \rightarrow H_2CO_3 \rightarrow H^+ + HCO_3^-$$

water+carbon dioxide → carbonic acid → hydrogen ions+bicarbonate

When the bicarbonate arrives in the lungs, it is reconverted to CO_2 and H_2O. From here the CO_2 crosses into the alveoli and is breathed out of the body.

$$\text{Lungs}$$
$$\uparrow$$
$$H^+ + HCO_3^- \rightarrow CO_2 + H_2O$$

Note that here and below I am giving a simplified overview of the role of bicarbonate in the transport of CO_2 in the blood. A more detailed and technical description is given in all major medical texts. The hydration of CO_2 to bicarbonate actually occurs inside the red blood cells in the tissues (catalyzed by the enzyme carbonic anhydrase), and the bicarbonate is re-converted to CO_2 and H_2O in the red cells as they pass through the lungs. For an excellent detailed description of the formation of bicarbonate, the transport of CO_2, and the various associated phenomena such as the chloride shift, see G. J. Arthurs and M. Sudhakar's 2005 article, "Carbon Dioxide Transport."[23]

There is something intriguing about the fact that another one of the end products of oxidation, water, generates bicarbonate by reacting with CO_2. Then because of water's fitness to serve as the medium for the circulatory system, water delivers the bicarbonate to the lungs where it reacts with hydrogen ions, reforming CO_2, which is exhaled from the body. Oxidation makes water, and water, as it were, makes bicarbonate, dissolves bicarbonate, and transports bicarbonate. Water is uniquely fit chemically to generate bicarbonate—by reacting with CO_2—and serves as the ideal medium of the circulatory system to transport the bicarbonate in the blood to the lungs.

So the solution to the problem of CO_2 excretion resides in the intrinsic chemical and physical properties of both water and CO_2, the two end products of oxidative metabolism. The solution to CO_2 excretion is clearly one of supreme elegance and parsimony.

Heat

Another waste product of oxidation is heat. Heat is a product of many chemical reactions, but oxidations produce more heat than almost all other reactions thanks to oxygen's unparalleled vigor. Heat, of course, is not entirely a waste product, as it enables warm-blooded animals to maintain their body temperatures well above their surroundings even when ambient temperatures are well below body temperature. Nevertheless, excess heat must be excreted from the body, and again, as in so many other cases reviewed in these pages, water comes to the rescue. As we've seen, its high specific heat ably soaks up excess heat, and its high heat conductivity eases its transfer from the tissues to the blood and from the blood to the skin, where water's high latent heat of evaporation (the highest of any molecular substance) has a powerful cooling effect, drawing excess heat out of the body.

ATP and Water

The other two products of the basic reaction need little further comment. ATP is the energy currency of life, the recipient of oxygen's chemical energy attenuated and transformed into the biologically useful form of high energy phosphate bonds. Water, the fourth product, is the very matrix of life. No product of any biochemical reaction could be more innocuous and at the same time serve a myriad of vital ends in the biological domain.

The Bicarbonate Buffer

AMAZINGLY, THERE is yet another remarkable consequence of water's reaction with CO_2 to form bicarbonate. It provides air breathers like ourselves with an ideal buffer—the bicarbonate buffer—for maintaining the blood's acid base balance.

The buffering works because the reaction:

$$H_2O + CO_2 \rightarrow \text{carbonic acid} \rightarrow \text{hydrogen ions} + \text{bicarbonate}$$

is readily reversible. This means that whenever the concentration of hydrogen ions in the blood rises, increasing acidity, as when lactic acid accumulates during anaerobic activity, such as dashing for a bus, the bicarbonate reacts with the hydrogen ions, generating carbonic acid that then dissociates into carbon dioxide and water. In other words, the reaction moves to the left:

Lungs

$\uparrow$ $\uparrow$

$H_2O + CO_2 \leftarrow$ carbonic acid $\leftarrow$ hydrogen ions + bicarbonate

$\leftarrow$

One reason for the excellence of the bicarbonate buffer system is that unlike an ordinary buffer functioning in a closed system, the CO_2 generated as H+ ion-concentration increases can be continuously eliminated from the body via the lungs. This greatly increases its efficiency over a buffer in a closed system. As Burton Rose points out in his *Clinical Physiology of Acid Base and Electrolyte Disorders*, calculations show that because of the ease with which the carbon dioxide (and with it, in effect, the hydrogen ions) can be breathed away, the buffering capacity of the bicarbonate system is increased ten to twenty times over that of an ordinary buffer working at its pH optimum.[24]

While the buffer functions in both air breathers and water breathers, the fitness of the bicarbonate buffer is greater for an air breather, because the level of bicarbonate available for soaking up excess hydrogen ions (acidity) is far greater in an air breather than in a water breather. This is because in an air breather the pCO_2 levels in the blood are determined by (in equilibrium with) the pCO_2 level of the air in the lungs, which is in intimate contact with the blood. This is about 44 mm Hg. This relatively high blood pCO_2 is a major determinant of the level of bicarbonate (HCO_3^-) in the blood, which in the case of humans at the normal pH of blood (7.4) is about 20–24 mmol/L.

But for fish, the blood is in contact with seawater, in which the pCO_2 is only 0.4 mm Hg,[25] far less than the pCO_2 of 44 mm Hg in the lungs of an air breather. So the pCO_2 in fish blood is only between 1–4 mm Hg,[26] and the level of bicarbonate available in a fish to soak up excess H ions and thereby defend against an increase in acidity is far lower (around 2–7 mmol/L) than that in a mammal.[27]

This means that in an air breather with a large reserve of bicarbonate and a high initial PCO_2 in the lungs, any increase in the acidity of the blood can be readily buffered by increased ventilation, which lowers the pCO_2 in the lungs, causing the bicarbonate to combine with the excess H ions, decreasing the level of acidity in the blood. However, the same ventilatory strategy is not available in a fish, as its reserve of bicarbonate is far lower and there is far less scope for ventilatory reduction in pCO_2 levels as blood pCO_2 is already low, close to that in the seawater in contact with the gills.

As James Claiborne and colleagues point out, "A primary mammalian response to metabolic acidosis is to increase lung respiratory exchanges in order to reduce the plasma partial pressure of CO_2 (pCO_2)." This increases pH (which means lowering the concentration of H ions in the blood). A high starting pCO_2 "allows the respiratory adjustments necessary to excrete more CO_2," drawing more acid out of the body. Fish, they further explain, cannot use this strategy to the same degree, due to the far lower pCO_2 and consequent much lower bicarbonate concentration typically found in these creatures:

> When compared to air breathers, the requirements for extracting sufficient oxygen from the water necessitate a 10- to 20-fold higher ventilatory water flow over the gills (normalized for blood oxygen capacity and flow rates). This water flow, combined with a high water CO_2 solubility, allows the rapid transfer of CO_2 across the gills into the water and reduces the PCO_2 of the plasma to only 1–4 mmHg in most species. At these low levels, additional respiratory reductions in PCO_2 are possible, but small. Thus, fishes must utilize another mechanism to excrete excess H^+.[28]

The other factor responsible for the high concentration of bicarbonate in the blood is the pKa of the bicarbonate buffer. The pKa of a buffer is the hydrogen ion concentration, or pH, at which the concentration of the acidic form (carbonic acid in the case of the bicarbonate buffer) and

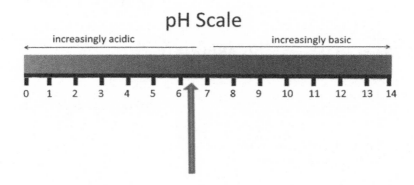

Figure 8.2. The pH scale. Arrow pointing to the pH of the bicarbonate buffer.

the basic form (bicarbonate) are equal. In the case of the bicarbonate buffer, this is about 6.4 in pure water and close to 6.1 in blood. Because a buffer functions optimally at a pH value close to its pKa level while the hydrogen ion concentration of blood is close to pH 7.4,[29] this would seem to be an anomaly. However, the calculation from the equation that relates the concentration of acid to base in a buffer at different hydrogen ion concentrations[30] shows that the ratio of bicarbonate to carbonic acid at pH 7.4 is about 20:1. This is a far higher level of bicarbonate than would be the case if the pKa were closer to 7.4. And this leads to a far higher level of bicarbonate to buffer against any increase in acidity of the blood than would be the case if the pKa were at 7.4. The pKa value of the bicarbonate buffer is just right.

Because of CO_2's volatility and how easily its levels can be regulated by alterations in ventilation, the bicarbonate buffer system provides an elegant mechanism for changing the blood's acidity level via changes in ventilation in an air breather. Many authors have commented on the fitness of the bicarbonate buffering system for maintaining acid-base homeostasis in air breathers such as ourselves. Like Henderson in the early twentieth century,[31] J. T. Edsall and J. Wyman were struck mid-century by the remarkable nature of the system. As they commented: "The com-

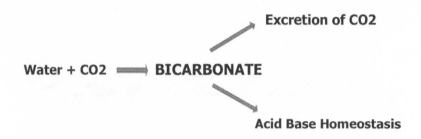

Figure 8.3. Perfect parsimony. Bicarbonate plays a crucial role in two vital physiological processes.

bination of the acidity and buffering power of H_2CO_3 with the volatility of CO_2 provides a mechanism of unrivalled efficiency for maintaining constancy of pH in systems which are constantly being supplied as living organisms are with acidic products of metabolism."[32]

The reaction of water with CO_2—a "waste product" of oxidation—in generating bicarbonate solves two basic and very different physiological problems: that of ridding the body of an end product of oxidative metabolism, and that of maintaining acid-base homeostasis. Thus both the problem of excreting a major end product of oxidative metabolism and the problem of acid/base balance are elegantly solved in the properties of the same remarkable compound. The solution is one of stunning elegance and parsimony, and one of particular benefit to air-breathing terrestrial organisms.

Saving the Best for Last

NATURE HAS left to the end of the vital process of oxidation what is perhaps one of the most remarkable of all ensembles of mutual fitness for air breathers. The facts are not in dispute. Two of the inevitable chemical end products of oxidative metabolism, water and CO_2, react together chemically to generate bicarbonate. The bicarbonate is used to transport CO_2 to the lungs, and water not only reacts chemically with CO_2 to generate HCO_3^- but physically transports bicarbonate to the lungs, a task that depends on the superb fitness of water to form the medium of the circulatory system (see Chapter 6). Moreover, the bicarbonate spontaneously generated by the reaction of water with CO_2 has ideal characteristics for buffering the blood of air-breathing organisms.

Is there an equivalent ensemble of fitness in all nature? (Perhaps only the teleological hierarchy of the successive ensembles of fitness for terrestrial life manifest in the hydrological cycle, described in Chapter 2, comes close.) The facts are undisputed and are described in every physiology textbook, and yet the wonder of how the chemical and physical properties of the players conspire so cleverly and parsimoniously to achieve the end of both CO_2 excretion and acid/base balance is seldom appreciated.

Whatever the reason, one thing is clear: this astonishing ensemble of environmental fitness has nothing to do with Darwinian selection. It was built into nature from the beginning, long before life arose and before Darwinian selection could act.

So now let's summarize the extraordinary mutual fitness of the main players involved in the excretion of the waste products of oxidation (water, CO_2 bicarbonate, heat) in air-breathing organisms like ourselves. (1) Oxygen and CO_2 are both gases in the ambient temperature range. This enables the uptake of oxygen and excretion of CO_2 to take place via the same route, i.e., the lungs. (2) Both oxygen and CO_2 are soluble in water and hence can be transported in the blood. (3) The two chemical end products of oxidative metabolism, water and CO_2, react together chemically to generate the bicarbonate radical (HCO_3^-) which has ideal characteristics for buffering body fluids of air-breathing organisms. (4) The bicarbonate radical (HCO_3^-) is used to transport CO_2 to the lungs. (5) Liquid water not only chemically reacts with CO_2 to generate bicarbonate, but water as the ideal medium of the circulation physically transports bicarbonate and dissolved CO_2 to the lungs. (6) By virtue of water's high heat capacity, it readily soaks up excess heat (the third end product of oxidative metabolism) and carries it to the periphery where, (7), its uniquely high latent heat of evaporation greatly aids the excretion of heat. That respiration in higher organisms should depend on such a profoundly beautiful and parsimonious synergy in the physical and chemical properties of the end products of oxidation beggars belief.[33]

In the context of such an extraordinary body of evidence, one is tempted to adapt astrophysicist Fred Hoyle's famous remark about the fine tuning of physics and chemistry for life: *A common sense interpretation of the facts suggests that a superintellect has monkeyed with the laws of chemistry and biology to enable the excretion of the end products of oxidative metabolism in air-breathing organisms like ourselves.*[34] Surely no one who reads the above paragraph and carefully considers the facts can reasonably deny that there is indeed a profound fitness in nature for beings of our biology—that we do indeed occupy a central teleological place in the natural order.

9. THE RIGHT PROPORTIONS

To keep us alive and living, the human body performs millions of complex functions throughout our lifetime. For example, in just 60 seconds, the human body takes 15 breaths, its heart beats 70 times, its tear ducts moisten the eyes 25 times, its brain conducts six million chemical reactions, its bone marrow produces 180 million blood cells, its skin sheds 10,000 particles of skin, and about 300 million of its cells die and/or are replaced. Furthermore, the human body manages to "extract the complex resources needed to survive, despite sharply varying conditions, while at the same time, filtering out a multiplicity of toxins."... The human body is a remarkable biological machine that is supported and maintained by well-structured and interdependent body systems and their unique organs.... It has about a hundred trillion cells, 60 miles of blood vessels, a 3-pound brain with 50 to 100 billion nerve cells and amazing thinking capacity, and 2.5 billion heart beats in a life time of 75 years, to name a few of its unique characteristics.

—Abour H. Cherif et al., "Redesigning Human Body Systems"[1]

W̶E SAW IN CHAPTER 5 THAT THE LUNGS IN HUMANS OCCUPY about 10 percent of the body's volume and that this volume suffices to provide us with the 250 ml of oxygen we require each minute to satisfy our energy needs. And we saw in Chapter 6 that the volume of blood sufficient to transport that 250 ml every minute to the tissues takes up about 9 percent of the body's volume. We also saw that these functional volumes result not only from sophisticated adaptation (witnessed in the marvel of the alveolar membrane) but are also constrained by a set of prior physical constants: viscosity of blood, pO_2 in the lungs, weight of air, solubility of oxygen, and so forth. And we saw that the value of these physical constants must be very close to what they are if the necessary functional volumes of these two organ systems are to be accommodated in the human body. Finally, we saw that if the environmental fine tuning of these constants had been less precise and the functioning of these two organ systems had thereby necessitated, say, twice the volumes they do, then not only our upright android design but indeed design of all other large terrestrial organisms would have been massively compromised. There would be, as mentioned earlier, citing Stephen Vogel, precious little room "left for guts and gonads,"[2] or brains or any other organ systems. Large mobile terrestrial organisms remotely comparable to ourselves would not exist.

The fitness of the functional volumes of our lungs and cardiovascular system is only the start. To be a large terrestrial carbon-based life form possessed of controlled mobility—a being that can explore and manipulate the environment—requires several additional organ systems, including the muscular, nervous, skeletal, and visual. And these organ systems must also be effectively accommodated and functionally integrated in the human body.

Here, too, nature obliges.

Muscles

EVEN A trained athlete finds it hard to lift significantly more than his own weight above his head. An ant, however, can easily lift many times its own weight, without training and seemingly without effort, and carry it over all manner of obstacles to the nest. Researchers at the University of Cambridge have photographed an Asian weaver ant lifting one hundred times its mass.[3] How is an ant proportionately so much stronger

Figure 9.1. Leafcutter ants, shown carrying loads greater than their own body weight.

than a trained human weightlifter? One might surmise that ant muscles are much stronger than our own. "When we see an ant carrying in its jaws a seed that weighs more than the animal itself, we gain the impression that its muscles must be inordinately strong," writes Knut Schmidt-Nielsen in his book *Scaling.* "However, measurements of insect muscles show that they are not stronger. In fact, they exert the same force per unit cross-sectional area as vertebrate muscles."[4] Moreover, the fine molecular structure of all muscles in all organisms, insects and vertebrates, is almost exactly the same.

But why, if the muscles have the same design and strength, is the ant able to lift so much more proportionate to body size? It's a matter of scaling. As size increases, the mass or weight of an organism increases by the cube of its length (L^3). But the force muscles can exert only increases by cross-sectional area, i.e., the square of length (L^2). Consequently, as Schmidt-Nielsen explains, "The force exerted by muscles, relative to mass, increases in proportion to the decrease in L. This is the reason that the ant appears to have muscles of unmatched strength."[5]

That force exerted by muscles relative to mass is inversely related to size imposes a severe limit on the size of all large terrestrial vertebrates and particularly on an upright bipedal organism like man. And it is because the strength of muscles relative to mass is inversely proportional to the organism's size that, as Rod Lakes comments, "A massive dinosaur will not be able to get up off the ground as easily as a squirrel. A human athlete who weighs 150 pounds can do pull-up exercises more easily than an athlete of similar body composition who weighs 300 pounds."[6]

The increasing weakness of muscles relative to mass as the size of an organism increases—the L^2/L^3 constraint—means that if the maximum power of muscles were even modestly less, then while an ant-sized organism might get along OK, the mobility and general design of organisms our size and larger would be enormously constrained, if not impossible. On any such decrease, a large dinosaur would in all probability not merely be unable to get off the ground as easily as a squirrel; it would be unable to get off the ground at all.

Determinants of Muscle Strength

IN ADDITION to the L^2/L^3 constraint, there are additional basic geometric and physical constraints on the absolute strength of muscles in all organisms. To begin with, the muscles of all organisms consist of densely packed arrays of the basic contractile elements. I wrote in *Nature's Destiny*:

> Each basic working component in the muscle cell is an individual protein molecule consisting of a long tail and short head rather like an elongated tadpole, known as a *myosin motor*. Movement comes about as a result of a sequence of three conformational changes. First, the myosin head attaches itself to another long fibrillar molecule known as actin... Second... the head bends suddenly—the power stroke—and this bending causes the myosin molecule and the actin to move in opposite directions. Third... the head unbends and attaches itself again to the actin [several nanometers further along the actin fiber]. The sequence is repeated again, and gradually, via a series of small steps, the two molecules slide past each other.[7]

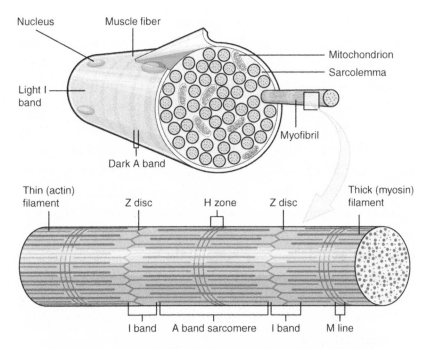

Figure 9.2. A muscle cell, showing the tight packing of the cell with myofibrils and the tight packing of the sliding elements in each myofibril.

As I further explained in that book, "each myosin head moves about eight nanometers along the actin fiber with each power stroke, and the heads are stacked in the muscle fibrils in a helical conformation about fourteen nanometers apart."[8] And from consideration of the geometrical constraints involved in packing molecular motors and actin filaments into muscle fibers, in all likelihood "no further improvement in muscle power can be achieved by increasing the density of packing of the myosin motors."[9] They are packed as tightly as possible. Schmidt-Nielsen writes that "all muscle contraction we know about is based on sliding filaments of actin and myosin, and if we could pack more filaments into a given cross-sectional area, the force would be increased." However, "this is most unlikely because the diameter of the filaments is determined by the size of protein molecules that make up the filaments and their size is probably determined by the requirement of the molecular mechanism."[10]

The absolute strength of muscles is also constrained by the strength of what are known as the weak chemical bonds (as distinguished from a different type of bond—the strong chemical bonds, in which atoms share electrons). These weak chemical bonds detach and re-attach the myosin head to the actin filament during each power stroke.[11] These types of chemical bond are involved not only in the reversible attachment of the myosin head to the actin in the muscle cell but also in the selective reversible binding of all the complementary molecular surfaces in the cell—including between the two strands of DNA, between two stretches of the polypeptide chain of a protein, and between an enzyme and its substrate. They impart to molecular surfaces what James Watson calls "selective stickiness,"[12] upon which the higher three-dimensional configuration and functioning of so much of the cell's vital activities, including the functioning of molecular motors, depends. Because of the many vital roles they play in the cell, they are, as Watson puts it, "indispensable to cellular existence."[13]

Moreover, as Watson also emphasizes, the energy of the weak bonds is just right for the varied cellular functions they perform. As things are, "enzyme substrate complexes can be both made and broken apart rapidly as a result of random thermal movement."[14] Problems would arise if the weak bonds were appreciably stronger or weaker. The amount of energy in these weak bonds is "not so large that rigid lattice arrangements develop within a cell—the interior of a cell never crystalizes as it would if the energy... were several times greater."[15] Conversely, the weak bonds are not so weak that it becomes impossible to fit a sufficient number on complementary surfaces to form a specifically shaped bond strong enough to withstand the hurly burly and continual thermal jiggling of the constituents within the cell. In such a counterfactual situation, reversible specific binding of a substrate to its catalytic site or a molecular motor head to an actin filament could never occur.

The reversible weak chemical bonds attach the heads of molecular motors to the actin filaments along which they crawl. The specific strength of these weak bonds determines the absolute strength of the individual power stroke of the molecular motors. And it is the force of the individual power stroke in conjunction with the geometric packing constraints which ultimately determine the strength of all muscles in all organisms. Again, as in so many instances, a biological end—in this case, muscle power—is determined by a set of prior physical environmental constraints.

The Right Proportion

GIVEN THE various physical constraints on absolute muscle strength, mammals must invest about 40 percent of their mass in muscle[16] to provide adequate mobility. (In passing, it is worth noting that the large muscle mass necessary for mobility has one seldom-mentioned advantage: the delicate nerves and blood vessels, by being embedded in the musculature, derive a significant measure of protection from injury.)

One thing is clear: if the prior fitness of physics and chemistry for musculature were such that muscles were appreciably less powerful than they are, major design problems would arise. For example, the strength of the grip of human fingers is generated by muscles in the forearm. The muscle bulk needed to provide the required grip strength for handling a number of vital tasks cannot be accommodated in the hand. The fact that it is necessary, even with the strength of muscles as they are, to place the muscles for grip in the forearm indicates the enormous difficulties one would face in attempting to redesign a mobile terrestrial organism of our size and mass if muscles were appreciably less powerful.

As in the case of the lungs and cardiovascular system, the necessary functional volume of muscle appropriate for the human body plan is highly constrained by a set of fine-tuned environmental constants. And so it seems our ability to move was prefigured in the order of things from the beginning.

Nerves

To BE mobile and fit for terrestrial life, we require not just muscles but finely controlled muscular activity. This necessitates fast reflexes, fast nerve conduction velocities, and thin nerve cords that take up only a fraction of the volume of the body. Fortuitously, nature obliges again.

Consider first conduction speed. Tim Taylor's description of the functioning of the nerves in our hands conveys the critical role of speed:

> The nerves of the arm and hand perform a substantial two-fold role: commanding the intricate movements of the arms all the way down to the dexterous fingers, while also receiving the vast sensory information supplied by the sensory nerves of the hands and fingers. The movements of the arms must be fast, precise, and strong to complete the diverse activities the body engages in throughout the day. Even the tiny hand muscles, which perform very delicate and precise movements, are driven by about 200,000 neurons. Rapid conduction of sensory nerve signals from the hands provides critical information to the brain and feedback during precise activities.[17]

As I wrote previously, "One area where very fast nerve conduction is vital is vision, more specifically, when keeping the eyes fixed on some object in the field of vision while in motion. With each step, the head moves and so do the eyes. If it were not for the speed of what is known as the vestibular-ocular reflex (VOR), vision in motion would present considerable difficulties,"[18] the result resembling, as some authors have described it, "a photograph taken with a shaky hand."[19]

Keeping our balance when we walk also requires rapid nerve conduction speed, since it necessitates continual second-by-second assessments of the position of the limbs in space and continual simultaneous coordinated contraction and relaxation of different muscle groups. Patients with the disease Hereditary Spastic Paraplegia, for example, have slow nerve conduction speeds due to degeneration of the peripheral afferent nerves that carry sensory information to the brain. This causes these persons to have great difficulty balancing and walking.[20] This unfortunate disease illustrates the fact that the rapid reflexes necessary for an organism to carry out finely coordinated motor activities are only possible when nerve conduction speed is very rapid.

The speed in different organisms varies over more than three orders of magnitude, from 10 centimeters per second in simple invertebrates to

120 meters per second in the nervous system of mammals.[21] The speed of nerve conduction imposes an absolute limit on the maximum size an animal can attain. No animal can be 100 meters long and at the same time nimble. Even at the fastest conduction speeds of 120 meters per second in a 100-meter long organism, a nerve impulse will take two seconds to travel from the brain to its extremities and back. Even an organism our size could never handle fire or undertake any sophisticated manipulation or exploration of the world if the maximum speed of nerve conduction were ten times less. Indeed, we would probably be unable to survive.

Compactness is another criterion that must be satisfied if nerves are to be fit to coordinate muscle activity in organisms like ourselves. Schmidt-Nielsen explains, noting that nerves carrying hundreds or even thousands of single axons (nerve fibers) control vertebrate muscles.[22] These nerve fibers are only five to twenty microns in diameter.[23] If for some reason they needed to be considerably thicker to attain the necessary speeds of 120 meters per second, Schmidt-Nielsen notes, this would necessitate "nerve trunks of inordinate size."[24]

From cable theory, which provides a quantitative description of current flow and voltage change, both within and between neurons,[25] it can be shown that conduction speed is proportional to the square root of axon diameter. In many invertebrates high conduction speed is gained by having axons of large diameter, and in the squid these can be one mm in diameter[26] (fifty times the diameter of fast-conducting axons in mammals).

So how do vertebrates get away with such small diameter nerve axons yet still achieve fast conduction speeds (even faster than the very large diameter high-speed invertebrate axons)? The optic nerve, for example, contains about one million nerve fibers carrying visual information from the retina to the brain.[27] But as Schmidt-Nielsen points out, despite this great number the optic nerve in humans has a diameter of only three millimeters. If it were to contain the same number of fibers the size of large invertebrate axons conducting at the same speed as those in the optic nerve, this would require a diameter of 300 millimeters (12 inches)—larger than the human head.[28] Similarly, the nerves servicing the muscles of the arm would have to be larger than the arm.

Vertebrate nerves are able attain such high conduction speeds with such very small diameter axons because of a crucial adaptive innovation. Rapidly conducting vertebrate axons are covered in a thin sheath of a fat-like substance, myelin, that "is interrupted at short intervals to expose the nerve membrane," writes Schmidt-Nielsen. These "exposed sites are known as nodes" and "are separated from one another by a fraction of a millimeter up to a few millimeters." This adaptation allows for what is termed "saltatory conduction," where the nerve impulse, instead of traveling sedately and continuously (and slowly) down the axon, jumps from node to node, vastly increasing transmission speed.[29]

Thus myelinated axons enable high conduction velocities without undue space occupied by the bundles of nerve fibers that make up the major nerve trunks.

Prior Fitness

While biological adaptation has played an important role in the evolution of myelinated nerves and saltatory conduction, as in so many other instances (such as the optimization of lung function discussed in Chapter 5) these innovations, indeed the very existence of nerve cells and nervous transmission, depend on several elements of prior natural environmental fitness.

One of the prime elements of fitness is the non-polar nature of hydrocarbon chains, which make up the fatty acid core of the cell membrane. Their nonpolar nature has two major consequences. First, it causes their spontaneous self-organization, driven by the hydrophobic force into a thin layer of material, which surrounds the entire cell—the cell membrane. Second, because of the nonpolar nature of the hydrocarbon chains of the cell membrane, the membrane is an electrical insulator. This allows for the generation of the membrane potential—that is, the difference in electrical charge between the outside and inside of the neuron, required for the subsequent generation of the nerve impulse.

Another element of prior fitness is the existence of small mobile ions—including sodium (Na) and potassium (K), which are ideally suited to move electric charge at great speed by diffusing rapidly down electrochemical gradients through the lipid bilayer via highly selective ion pores—referred to as ion channels or gates. What makes sodium and potassium ideal for this work is that they bind only weakly to organic compounds, rendering their mobility high.[30] Indeed, their mobility and capacity to move rapidly down gradients through ionic channels is astounding. As Bruce Alberts and his colleagues note, "Up to 100 million ions may pass through one open channel each second."[31]

Without nature's provision of these highly mobile inorganic ions, no cell would be able to regulate or generate a membrane potential or generate a nerve impulse. No other small particles of matter possess charge plus great mobility.[32] Neither proteins nor any of the organic molecules in the cell have the right properties to stand in for the alkali metal ions.

That one of the basic properties of the cell membrane, its electrical insulating character and the membrane potential it enables, should provide precisely the electrical characteristics required for the transmission of electrical impulses between cells and ultimately for the construction of the nervous system of beings like ourselves, is surely of great significance, providing further support for the basic theme of this book, that nature's fitness is not just for the generic carbon-based cell but, beyond this, for advanced multicellular organisms like ourselves.

In the properties of the mobile inorganic ions and in the basic structure and properties of the cell membrane, nature, it seems, knew we were coming.

The Brain

THERE IS of, course, one vital organ that is part of the nervous system and every bit as essential to our basic body plan and physiological functioning as lungs, blood, muscle, nerve, and bone—the brain. No other biological brain on earth comes close to the cognitive ability of the human brain, and no AI, despite the enormous efforts made in the field of artificial intelligence over the past two decades, is remotely comparable.

The sizes of animal brains vary enormously. As Ursula Dicke and Gerhard Roth note, "Among mammals, the smallest brain is found in the bat, *Tylonycteris pachypus*, which weighs 74 mg in the adult animal, and the largest brains of all animals are found in the sperm whale (*Physeter macrocephalus*) and killer whale (*Orcinus orca*), with up to 10 kg. African elephant brains weigh up to 6 kg."[33]

However although whale and elephant brains are bigger than the human brain, the total number of neurons in the neocortex (the outer layer of the cerebral cortex, the information-processing region of the brain and the seat of general intelligence) is less in cetaceans and elephants than in humans. Dicke and Roth comment:

> Owing to their large cortex volumes, their small neurons and high NPD [neuron packing density], primates have many more cortical neurons than expected on the basis of absolute brain size. The relatively small New World squirrel monkey has 430 million, and the much larger Old World rhesus monkey about 480 million, the New World white-fronted capuchin 610 million, gorillas 4300 million, chimpanzees about 6200 million, and humans about 15,000 million cortical neurons. The largest number of cortical neurons in non-primate mammals is found in the false killer whale with 10,500 million and the African elephant with 11,000 million, which is less than the number found in humans, despite the much larger brains of the former two.[34]

However, the number of neurons is not in itself the best guide to what Dicke and Roth call "general intelligence"—or more formally "information processing capacity" (IPC). As they point out, "Besides the number of cortical neurons and synapses, another factor that is important for cortical IPC is processing speed, which in turn critically depends on (i) interneuronal distance, (ii) axonal conduction velocity and (iii) synaptic transmission speed. Interneuronal distance is determined by NPD [neural packing density]: the higher the NPD, the shorter, trivially, is the interneuronal distance."[35] Taking these factors into account they conclude that "the highest IPC is found in humans, followed by the great apes, Old World and New World monkeys.... The IPC of cetaceans and elephants is much lower because of a thin cortex, low neuron packing density and low axonal conduction velocity."[36]

Michel Hofman, a researcher at the Netherlands Institute of Neuroscience, waxes lyrical, describing the human brain as "one of the most complex and metabolically efficient structures in the animated universe."[37]

But the human brain may be something more. Many authors have concluded that it may be very nearly the most intelligent/advanced biological brain possible. That is, its information-processing capacity may be close to the maximum of any brain built on biological principles, made of neurons, axons, synapses, dendrites, etc., and nourished by glial cells and provided with oxygen via circulation. For example, Peter Cochrane and his colleagues, in a widely cited paper, conclude "that the brain of *Homo sapiens* is within 10–20% of its absolute maximum before we suffer anatomical and/or mental deficiencies and disabilities. We can also conclude that the gains from any future drug enhancements and/or genetic modification will be minimal."[38] Hofman concurs: "We are beginning to understand the geometric, biophysical, and energy constraints that have governed the evolution of these neuronal networks. In this review, some of the design principles and operational modes will be explored that underlie the information processing capacity of the cerebral cortex in primates, and it will be argued that with the evolution of the human brain we have nearly reached the limits of biological intelligence."[39]

This conclusion is remarkable for two reasons. First, it means that what may be the most advanced brain in the "animated universe," close to the limits of biological intelligence, has a volume of only about 1.5 liters and thus can be conveniently placed within the skull atop the upright android form of a human. If near maximal information-processing speed had necessitated a brain two or three times as large as the average human brain, reengineering the human form to support such an outsized brain would have posed enormous problems (analogous to the problem discussed in Chapter 5 of fitting an oversized lung into the human body).

Second, because the energy demands of neural tissue are very high (the brain takes up 20 percent of the metabolic energy of the body), the maximum size of a brain that could be supplied with sufficient oxygen is also very close to 1.5 liters. A brain only twice the size of ours would take upwards of 40 percent of the lungs' oxygen uptake, leaving too little to supply the other organs of the body.

In sum, it is surely further evidence of nature's fine tuning for the human body plan that such a high-performance brain need only occupy a relatively small fraction of the body's volume, a volume commensurate with our upright bipedal stance, an android form of the right size and proportion to control fire (see Chapter 11), the great technological breakthrough which led humankind via metallurgy to the advanced technology of the twenty-first century. How fortunate that the size of what may be close to the smartest possible biological brain—capable of the miracle of understanding the world and transforming it through technology and culture—is perfectly commensurate with the design and dimensions of the human frame and body plan. Here is a fortuity wonderfully consistent with the anthropocentric conception of nature, of a natural order uniquely fit for beings of our biological design.

Ultimate Complexity

THE HUMAN brain might be something even more extraordinary still. It may be not just the most complex biological thinking organ possible, but close to the most complex functional organization of matter possible. Many scientists and science writers have waxed lyrical on contemplating the staggering compacted complexity of the human brain. As the author of a *Nature News and Views* article commented some time ago regarding brain neurons, "The latest work on information processing and storage at the single-cell level reveals previously unimagined complexity and dynamism. We are left with a feeling of awe for the amazing complexity found in nature. Loops within loops across many temporal and spatial scales."[40]

Each cubic millimeter of human cerebral cortex (the thinking part of the brain) contains 15,000 neurons,[41] 15,000 glial cells[42] (which provide support for the neurons[43]), some four kilometers of axonal wiring, 500 meters of dendrites,[44] and 400 million synaptic connections (the highest recorded density of synapses in any mammalian brain[45]). Contrast this with the number of components in a 747 jetliner, which is often cited to be about six million.[46] That's sixty times less that the synaptic connections in one cubic mm of the brain. Altogether the human brain performs on the order of an unimaginable 10^{15} synaptic operations per second.

My hunch is that it will turn out that the human brain does represent very close to the most complex functional assemblage of matter possible in our universe. Whether or not this does turn out to be the case, the evidence currently available is consistent with such a conclusion. We should be filled with a sense of awe.

Bones

FOR THE muscular and nervous systems to successfully initiate and control movement in large terrestrial organisms like ourselves, another organ system is needed, a structurally rigid endoskeleton. This consists in all higher vertebrates of a set of strong rigid elements, bones, linked together in a complex articulated structure by tendons and connective tissue.[47] The *Encyclopedia Britannica* ably summarizes many of the needs met by such a skeleton. The endoskeleton provides "structural support for the mechanical action of soft tissues, such as the contraction of muscles and the expansion of lungs," affords "protection of soft organs and tissues, as by the skull," and provides both "a protective site for specialized tissues such as the blood-forming system (bone marrow), and a mineral reservoir, whereby the endocrine system regulates the level of calcium and phosphate in the circulating body fluids."[48]

The article also notes that the endoskeleton is of greater import to terrestrial organisms than to fish: "Although a rigid endoskeleton performs obvious body supportive functions for land-living vertebrates, it

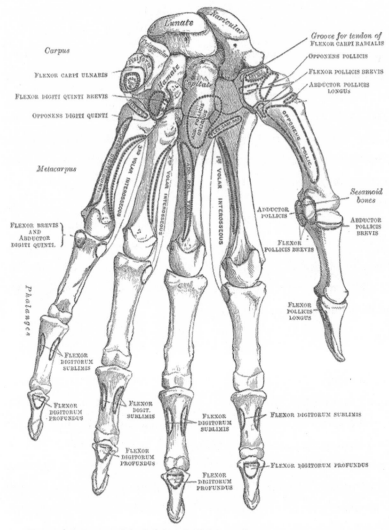

Figure 9.3. Bones of the left hand. Volar surface.

is doubtful that bone offered any such mechanical advantage to... bony fish... for in a supporting aquatic environment great structural rigidity is not essential for maintaining body configuration."[49]

Bone is a composite substance made up of an organic component which contains collagen fibers and a mineral component containing calcium phosphate called hydroxyapatite—$Ca_5(PO_4)_3OH$.[50] The crystals of hydroxyapatite confer compressive strength, and the collagen fibers in intimate association with the calcium phosphate crystals confer tensile strength and elasticity. Although rigid and able to sustain compressive forces and resist deformation,[51] bone is living tissue and capable of being continually remodeled, as occurs during growth and development in all higher vertebrates and in response to various environmental pressures.

Although bone is a complex biological adaptation, again, as in so many instances, its adaptive properties are enabled by prior elements of fitness in nature. For example, there is no obvious alternative to the major mineral component of bones, hydroxyapatite, which is thermodynamically the most stable calcium phosphate compound under physiological conditions. Moreover, in 500 million years of vertebrate evolution, no group has replaced the hydroxyapatite in bone with any other mineral component.

Prior fitness also means that in humans the volume of the skeleton need take up only about 11 percent of the body's volume.[52] This is yet another fortuity, for if the volume needed to afford a suitable strong and rigid set of articulated elements—the endoskeleton—had necessarily comprised significantly more than 11 percent of the body, then our motile lifestyle (and that of other motile large terrestrial vertebrates) would have been greatly constrained.[53]

The Eye

FINALLY, WHAT is perhaps the most dramatic example of an organ with a volume fortuitously commensurate with our bodily design is the eye.

Vision is not the only sense capable of providing a detailed image of the environment. An alternative is echolocation. Bats use it to successfully navigate through thick vegetation in pitch darkness to hunt and catch flying insects. Even some humans who are blind can learn to master the technique, in at least one case enabling the person to play football, basketball, and even ride a bicycle through traffic.[54] Even the sense of touch shared by all organisms—taken to extraordinary length in the case of the star-nosed mole, which has 25,000 touch sensors on

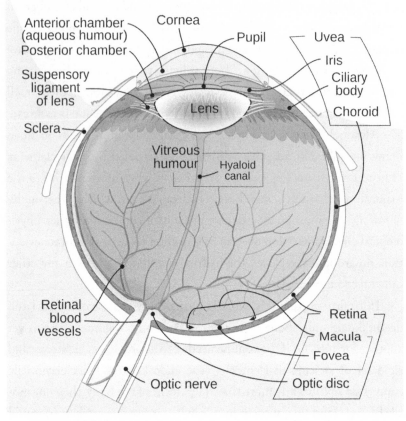

Figure 9.4. The human eye.

its "nose," many more in an area the size of a human fingertip than in the whole human hand[55]—can bring much essential information about one's immediate environment. And in some organisms, such as dogs and bears, the sense of smell is very highly developed.

It's true that sensing the environment using echolocation may enable many organisms to navigate with extraordinary efficiency, and a great deal of information regarding our immediate environment can be gained by the sense of touch or even smell. But no organism restricted to the use of sound, touch, or smell would ever "see" the moon and stars, witness a lightning strike, observe a *Paramecium* down a microscope, or watch the dancing flames of a campfire. And it is very doubtful that organisms restricted to sensing their environment through sound, touch, taste, and smell would ever master fire. Fire would remain a mysterious and frightening fiery demon to any eyeless creature, however intelligent, something beyond control. And thus the route to metallurgy, chemistry, and advanced technology would remain foreclosed forever. Moreover, it was seeing the stars and planets, and recording their regularities, that spurred the development of science.[56] High acuity vision is thus essential for any species capable of exploring in detail its surrounding environment and developing a technology.

The text below reviews the remarkable prior environmental fitness for high acuity vision in an organism of approximately our size and design.

The Right Energy: All biological light-detecting devices depend on the fundamental fact that the energy levels of electromagnetic radiation in the visual region are just right for photochemistry and bio-detection. As noted in Chapter 3, visible light is the only type of electromagnetic radiation with the appropriate energy level for interaction with and detection by biological systems. Ultraviolet, X-ray, and gamma ray photons are too energetic and highly destructive, while photons in the infrared, microwave, and radio regions lack the energy necessary to activate biomolecules for specific chemical reactions and for detection by biosystems. The Goldilocks visual region represents the tiniest fraction of the entire span of the electromagnetic spectrum—equivalent to one playing card in a stack of cards reaching beyond the Andromeda galaxy.[57]

The Right Wavelength: But the fact that electromagnetic radiation in the tiny visual region has the right energy levels for bio-detection, while necessary for vision, is not sufficient. The wavelength must also be right for the formation of an image in a high acuity eye fit for biological organisms of our size and design.

Like ocean waves passing through the entrance to a harbor, when light waves pass through an opening or aperture—the pin hole of the *Nautilus*, the pupil of the human eye, the aperture of a telescope or camera—they suffer dispersion, also called diffraction. That is, they bend

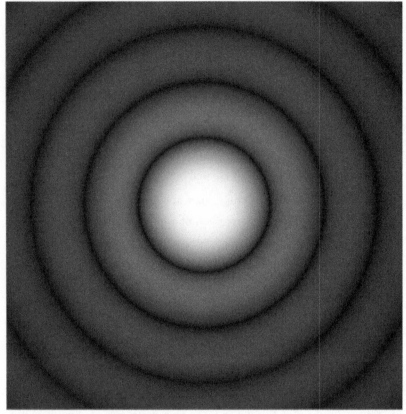

Figure 9.5. A computer-generated Airy disc.

outwards after passing through the aperture. One consequence, first observed by astronomers, is that when light from a point source (a star) is
focused to an image in a telescope, instead of being focused to a point it
spreads out into a tiny disc surrounded by concentric rings, the so-called
Airy disc, named after Georg Biddell Airy (who first provided an explanation for the phenomenon).[58]

The formation of the disc, whether in the eye or telescope, reduces
the resolving power of the optical device because when two point sources
in the visual field are close together, their Airy discs may overlap, and the
two sources cannot be resolved. For this reason the image resolution of
all optical devices that focus an image through an aperture is said to be

diffraction limited, an inevitable consequence of the wave-like nature of light.

The formal relationship in an eye or camera between disc diameter (degree of diffraction), wavelength, aperture, and focal length is given in the formula derived from Airy: Airy disc diameter $= 2.44 \cdot \lambda \cdot F/A$ (where λ is wavelength, F is focal length, and A is aperture). And note that from the formula, disc diameter (or degree of diffraction) is directly proportional to λ (wavelength) and inversely proportional to the diameter of the aperture.[59]

From this formula, if focal length (F) is 1.7 cm (as with the human eye), and if the aperture (A) is 0.8 cm (maximum pupil diameter of the human eye), then the ratio F/A is close to 2. And if the wavelength (λ) is 0.5 microns (the middle of the visual band), this gives an Airy disc of close to 2.5 microns in diameter. This is the minimum disc size achievable on the human retina and is therefore the maximum resolving power of the human eye. These figures suggest that the size of the Airy disc on the human retina is close to the minimum diameter of the photoreceptors.

Just like an ocean swell entering a harbor, where the degree of diffraction of the waves is directly proportional to the wavelength of the swell (the distance between successive waves), in the case of light, the larger the wavelength the larger the diffractional effect, the larger the Airy disc, and the poorer the resolution.

And also just like ocean waves entering a harbor, where the bigger the entrance the less the waves bend as they enter, the bigger the pupil of the eye or the aperture of a camera, the smaller the diffractional effect, the smaller the Airy disc, and the greater the resolving power of the device.

What might the consequences be if the wavelength of light suitable for photochemistry and bio-detection had been ten times greater? The only way to compensate for the increased diffraction (Airy disc diameter) and decreased resolution would be to increase the aperture's diameter tenfold, necessitating an eye the size of the head. And if the wavelength of light were one hundred times greater, then rescuing resolution would necessitate an eye of diameter two meters, the size of a golf cart. Conversely, if wavelength for photochemistry and bio-detection were significantly less, diffraction would be less and the Airy discs on the retina smaller. However, no advantage in terms of resolution could be gained because the smallest feasible photodetector cells would have a diameter larger than the Airy diffraction discs. (For a more detailed and comprehensive explanation, see Chapter 5 in my earlier work, *Children of Light*.)

Amazingly, it turns out that the wavelength of light is fit for both bio-detection and just right for a high acuity eye in organisms of our approximate size and biological design. And for this stunning coincidence we have physics, not Darwinian natural selection, to thank.

Plato's demiurge got it exactly right again: Given the immense range of wavelengths in the electromagnetic spectrum, from a fraction of a nanometer in the gamma region to tens of kilometers in the radio regions—a range of more than 10^{25} (see Chapter 3)—only light has the right energy levels and the right wavelength for high acuity vision in an optical device the size of the human eye. This is surely one of the most dramatic examples of fitness for our being in all nature, and I am sure that many readers will see this as strongly suggestive of design.

A Privileged Place

THE EVIDENCE presented in this chapter provides further confirmation that there is indeed a special fitness in nature for large mobile terrestrial carbon-based life forms very close in anatomical and physiological design to modern humans, that the laws of nature are indeed anthropocentric, and that beings of our biological design do indeed occupy a very special and privileged place in nature.

10. Fire and Metal

It seems inevitable that rocky planets, like bakers, cannot resist making crusts, heat being the prime cause in both cases. Although trivial in volume relative to their parent planets, crusts often contain a major fraction of the budget of elements such as heat-producing elements potassium, uranium and thorium as well as many other rare elements, while the familiar continental crust of the earth on which most of us live is of unique importance to *Homo sapiens*. It was on this platform that the later stages of evolution occurred and so has enabled this enquiry to proceed.

—Stuart Ross Taylor and Scott M. McLennan, *Planetary Crusts*[1]

The ancient Greeks, who had an answer to most things, believed that Prometheus brought down fire from heaven—and got himself into much trouble with Zeus for doing so. "From bright fire," says Aeschylus in *Prometheus Vinctus*, "they will learn many arts."

—A. J. Wilson, *The Living Rock*[2]

IN 2011 A CAR-SIZED, NUCLEAR-POWERED ROBOT NAMED CURIOSITY was launched from Cape Canaveral, Florida, and landed on the surface of Mars the following year, sent there to search for signs of life by employing a tiny automated laboratory built to detect organic chemicals and water. As of February 2022 Curiosity is still functioning without any hands-on assistance from its NASA creators, who are millions of

Figure 10.1. The NASA rover Curiosity on the surface of Mars.

miles away busily decoding and analyzing the cryptic messages it beams back to Earth. Curiosity has been followed by Perseverance and the first Chinese Mars lander, both dedicated to the search for life in the soils of Mars. These landers are just three representatives of a menagerie of present-day robotic marvels. Other engineers are crafting and testing smaller, more sophisticated robots, including drones the size of a bee for surveillance, search and rescue, and military operations.

These marvels are only possible because of the twentieth-century development of the computer and the astonishing miniaturization over the past few decades of microprocessors and other computer components. A 2008 article describes what was then a cutting-edge Intel microchip, one containing 47 million transistors, so tiny that two million of them could sit on the period at the end of this sentence, and which switch on and off up to 300 billion times per second. The author likens the process to "planning a city so tiny it could fit into a single bacterium."[3] In the intervening years since that article was written, that chip has been superseded by ones far faster and more compact. As I wrote in *Fire-Maker*: "After only a few centuries of science and only 200 generations since the first metal tool was manufactured, modern technology has reached the stage when its accomplishments increasingly resemble what would have seemed to our ancestors a form of magic."[4]

More such magic may be on the horizon, including the construction of microminiaturized robots far smaller even than a bee-sized drone, and the development of quantum computing, which may increase the memory and power of computers by many orders of magnitude. Fifty years ago the physicist Richard Feynman speculated about the possibility of building nanomachines by directly manipulating atoms,[5] an idea more recently popularized by futurologist Eric Drexler, who envisaged assembling tiny nanomachines tailored to carry out a host of functions, including a miniature robot submarine able to navigate the smallest capillaries in the body and designed to carry out various medical tasks, including seeking out and destroying individual cancer cells; rebuilding, atom by atom, damaged molecules; and repairing DNA.[6]

As I pointed out previously:

The dramatic technological advances over the past 100 years have provided extraordinary devices which have enabled mankind to gain enormous knowledge of the natural world, from the structure of the cosmos to the structure of DNA, more than in all previous centuries. Using light and radio telescopes we have peered at distant galaxies billions of light years from earth and looked back to the beginning of time, to the fire ball in which our universe was born. We have estimated the age of the universe and determined its dimensions. We have detected other worlds orbiting distant stars and estimated the number in our galaxy alone to be in the order of tens of billions! We have discovered how atoms are synthesized in the stars.[7]

Applying other technologies we have accelerated subatomic particles to near the speed of light and smashed them into the nuclei of atoms to reveal the hidden realm of the unimaginably small within. We have learned of the paradoxes of the quantum realm—of wave particle duality and the weirdness of entanglement. We have learned that the continents move over the surface of the earth, measured their stately progressions at rates of about one centimeter a year, and divined the shape and position of the continents as they existed hundreds of millions years ago. Using other clever devices we have determined the spatial positions of thousands of atoms in the basic macromolecules of life, measured the electrical activity of individual neurons in the living brain, and witnessed the electrical changes which accompany thoughts. With mini-cams we have looked inside the blood vessels of a beating heart. Through our machines we have achieved a miracle of scientific understanding in an instant of geological time.

And it is of course this vast increase in knowledge which makes possible the defense of the anthropocentric thesis presented in these pages.

Fire

OUR SCIENTIFICALLY and technologically advanced civilization of today grew out of a long series of hard-won discoveries and technological developments over many thousands of years, but of all the discoveries made in the course of humankind's long march, there was one that served as a fountainhead for a great deal of the advances that followed. Humankind discovered how to make and tame fire. Charles Darwin rightly saw the discovery as "probably the greatest, excepting language, ever made by man."[8]

We may never know exactly how or when our ancestors first mastered fire. It may be that several different groups of early humans did so independently. But we do know that our Pleistocene ancestors would be very familiar with fire. John Gowlett explains:

> Without doubt, natural fire was available on the landscapes inhabited by hominins. Of the millions of lightning strikes that are recorded each year, many lead to bush and forest fires, especially at the start of a rainy season: then lightning from the first thunder storms often strikes when much of the vegetation remains dry. Most of the instances of relevance are in forest, woodland and savanna, but the fire regimes operate surprisingly far north. Farukh and Hayasaka give the example of Alaska, where up to 100 fires are burning on a given day in the summer season, and important for hominins, they have an average duration of more than 20 days.[9]

Much of the *when, why,* and *how* remains to be unraveled, says Gowlett, but we do have clear evidence of burning "on numbers of archaeological sites from about 1.5 Ma onwards (there is evidence of actual hearths from around 0.7 to 0.4 Ma); that more elaborate technologies existed from around half a million years ago, and that these came to employ adhesives that require preparation by fire."[10]

Perhaps humankind's first use of fire was fire foraging—exploring burnt-out areas to retrieve various edible matter left by the retreating flames. But while fire foraging may have been our first use of fire, the first step towards actual control of fire was probably keeping a natural fire, sparked perhaps by a lightning strike, alight for a period of time. This would necessitate collection of wood to keep the embers alive as well as knowledge of slow-burning material like dung.[11] An early next step, and a monumental one, would have been the invention of ways to initiate a fire—such as rubbing a stick in a groove like Tom Hanks does in *Castaway*, or by using a simple hand-driven drill as hunter gatherers of the Kalahari employ (see Figure 11.1). We know that by 120,000 years ago, some groups possessed twine and leather cord, which would have been necessary to operating a bow drill.[12]

Metallurgy

ONE OF the first technological advances which followed the mastery and use of fire was the discovery that clay can be molded and then hardened via fire into various-shaped containers, initiating what was humankind's first industry, ceramics.[13] This in turn led to the construction of special kilns for firing and glazing different types of pottery under controlled conditions. And this subsequently led to the discovery that the high temperatures reached in a pottery kiln could be used for extracting and smelting metals from their ores.[14]

There is evidence that humankind mastered copper smelting as early as 7,000 years ago.[15] The subsequent extraction of copper from copper-bearing ores and its mixture with tin to make bronze (which is harder than copper) was independently discovered by cultures in both the Old and New Worlds, and ushered in the Bronze Age in the ancient Near East about 3500 BC.

The iron age was still to come. Copper smelting requires temperatures between 1,150°C and 1,250°C, but the smelting of iron requires even higher temperatures,[16] more advanced techniques, and more sophisticated kilns.[17] Consequently, iron smelting wasn't mastered until about 1200 BC, initiating the Iron Age. Iron was coveted for its greater hardness over bronze, making it a superior metal for weapons of war as well as for other purposes—including the manufacture of all manner of strong and durable tools, from ploughshares to needles.

Another major discovery, which played an important role in the development of metallurgy, was the discovery and manufacture of charcoal (formed by cooking wood in anoxic conditions). Burning charcoal in a kiln generates far higher temperatures than burning uncooked wood,[18] making it far easier to extract metals from their ores and rendering the fire hot enough to smelt and extract the most useful and important of all metals, iron.[19] This was a landmark event, which because of the ready

Figure 10.2. Blast furnace light in the town of Coalbrookdale, England.

availability of iron ores throughout the world, and because of the great utility of iron and its alloys, led to the spread of iron metallurgy, and the use of iron tools, extensively throughout the old world.

The importance of metallurgy's advent can hardly be exaggerated. As Arthur Wilson comments, "In whatever manner the secret of metallurgy was unraveled—and we shall never know precisely—it was a momentous step along the road to civilization… man, though still stumbling, entered the Age of Metals and opened up undreamed of possibilities for his future."[20]

As one author remarked, "Today it would be hard to imagine our civilization without metals. Just about every manufactured object we have either has metal in it or was made by metal machines and transported on ships, trains, or trucks made of metal. Without metals, we would literally be living back in the Stone Age."[21] It's true that some cultures have achieved some amazing things without metallurgy—the Maya, for instance—but it's highly doubtful than any beings anywhere in the cosmos could develop advanced technology remotely close to ours without metallurgy.

Metallurgy is only one of a host of other fire-assisted technologies. As Stephen Pyne noted in his *Vestal Fire*, "In fact almost no device or pursuit has lacked an element of combustion technology.... Fire distilled seawater into salt, wood into tar, resin into pitch and turpentine, grain and grape into alcohol; it transformed wood into ash and then into soap, and cooked calcitic rock into lime."[22]

Human determination, inventiveness, and sheer genius unquestionably played a central role in the development of metallurgy and other fire-related technologies, but human ingenuity is only part of the story. On even the most cursory reading, the control of fire, the development of metallurgy, and many of the subsequent technological advances that these discoveries enabled were only possible because of an outrageously fortuitous set of environmental conditions.

As we saw in Chapter 4, there are several elements of fitness in nature which enabled humans to utilize fire. To begin, there is the remarkable element of fitness that the same oxygen level in the atmosphere supports both human respiration and combustion. Then there is the unique attenuation of the reactivity of oxygen at ambient temperatures, which prevents the spontaneous combustion of the reduced carbon compounds in the biosphere (including in the human body), an attenuation enhanced by the relative non-reactivity of the carbon atom. We have all experienced this attenuation in the difficulty of starting a camp fire, especially in cold and wet conditions. Because of the curious non-reactivity of the oxygen at ambient temperatures, oxygen must be activated in the body for chemical reaction by special enzymes, but here the fitness of nature comes to the rescue, providing the transition metals with their unique properties to exploit oxygen's energetic potential in the body.[23] Then there is the existence of a powerful diluent in the atmosphere—nitrogen—which retards the speed of the spread of fire, rendering fire controllable and safe. It is only because of these various elements of prior fitness in nature that our ancestors could both exist and master fire, a mastery that served as the first and crucial step on the road to metallurgy and subsequent technological advances. But these elements of fitness for mastering fire and metallurgy are only the beginning. What follows are several more.

Wood

A VITAL resource for fire-making is wood, which is provided by the trunk and large branches of trees. Hot sustainable fires cannot be fueled by grasses or twigs. And only by using relatively large logs can a hot fire of the sort needed to fire and glaze pottery in a kiln be initiated and sustained. Such wood is also required for fires hot enough for metallurgy.

We saw in Chapter 2 that trees themselves require several vital elements of prior fitness in nature, including water-retaining soils, produced as a result of the weathering and erosion of the rocks in the course of the hydrological cycle, itself possible thanks only to several uniquely fit properties of water. Trees also depend on another ensemble of fitness in the properties of water, which enables the rise of water from their roots to the leaves at the top. This ensemble[24] includes the very high surface tension of water, which draws water up the narrow conduits in the stems of plants and trunks of trees, and a little understood and counterintuitive property of water, its remarkable tensile strength,[25] the result of its unique hydrogen-bonded network. Its tensile strength enables an unbroken, continuous column of water in the stem conduits of plants and tree trunks, an essential prerequisite if water is to be drawn continuously up from root to leaf. Vogel waxes lyrical about the overall mechanism, calling it a tale "mirabile dictu" (i.e., wonderful to relate).[26]

Photosynthesis

THE OXYGEN in the atmosphere and the reduced carbon fuels for fire-making are both the result of photosynthesis. And this unique process itself depends on several vital elements of prior fitness in nature (see Chapter 3). Recall that using the sun's radiant output to power photosynthesis is only possible because of the fact that the photons of visible light radiated by the sun have exactly the right energy levels to promote biochemical reactions, as well as the fact that the atmosphere is almost completely transparent to visual radiation, allowing it to penetrate to the

Earth's surface and be used by the photosynthetic apparatus of green plants.

Charcoal

IT IS hard to achieve temperatures sufficiently hot to smelt metals from their ores, particularly iron, using uncooked raw wood as a fuel. To generate temperatures sufficiently hot requires the use of coal or charcoal as a fuel.[27] Charcoal, unlike uncooked wood, is also an excellent reducing agent, greatly assisting the stripping of the oxygen from metal ores.[28] So charcoal provides both high temperatures *and* reducing conditions, precisely what's needed to extract metal from its ores in the kiln.

Arthur Wilson singles this out as "a fortunate coincidence."[29] Indeed, the fact that charcoal can be readily made from wood, and that a charcoal fire can reach sufficiently high temperatures *and* provide the reducing conditions in a kiln to enable the smelting of iron—the advance upon which so many subsequent technological advances depended—represents a remarkable element of fitness in nature for metallurgy.

The Right Temperature

ANOTHER WAY nature lent a hand in the development of metallurgy and the use of metals is that at ambient temperatures, metals such as copper and iron are solids (not liquids or gases) and possess high tensile strength.[30] They can be shaped into all manner of tools and structures at higher temperatures—into everything from steel beams for skyscrapers to delicate medical implements—and then hardened as they return to ambient temperatures. The ambient range is again a Goldilocks zone. Metals soften when temperatures move well above ambient temperatures. (Even steel loses tensile strength above 400°C.) And at much below zero, many metals become increasingly brittle.[31] But in the ambient temperature range, they are "just right," the same temperature range which is fit for carbon-based life.

Electricity

METALS ARE also excellent conductors of electricity, and because they are also ductile, they can be stretched into strong, thin wires. This fortuitous combination of conductivity and ductility made possible the construction of electric motors and generators, on which our technological civilization depends.

This dual fitness of metals for electrical technology is an arresting fact, but there's more. Several metals—especially copper, the conductor *par excellence*—are far better conductors at ambient temperatures than at higher temperatures. Copper, for example, conducts electricity ten times more efficiently at 100°C than at 600°C.[32] If the conductivity of copper were ten times less in the ambient temperature range, wires would have to be ten times the cross-sectional area to provide the same conductivity, ruling out many applications and making the construction of motors and dynamos far more difficult.

Copper is also unusually well-suited for electricians to work with, offering a unique combination of flexibility, hardness, and tensile strength, allowing electricians to twist strands of it together, pull it without breaking it, and cut it without easily crimping or nicking it. And its oxidized form also conducts electricity, making copper connections less prone to overheating than it would otherwise be, contributing to its longevity as functional electrical wire.[33]

Magnetism

ANOTHER VITAL property of many metals for electrical applications is magnetism, which plays a crucial role in electric motors and dynamos. The driving force in an electric motor arises from the attraction of the magnet in the central rotating armature with the stationary magnets in the stator, which surrounds the central armature. Electric motors are found in devices as diverse as industrial fans, blowers and pumps, machine tools, household appliances, power tools, disk drives, motor vehicles, electric watches, and space ships. So ubiquitous and vital is the electric motor that alongside the microprocessor, it is one of the defining devices of modern civilization.

The functioning of electric motors and dynamos testifies not only to the unique fitness of the metals for our electric age but also stands as another wonderful example of how unique properties of particular elements work together to serve a vital technological end. For motors and dynamos, the iron might be replaced by nickel (another magnetic metal), and the copper by silver (another excellent metallic conductor of electricity), but for dynamos and motors, no non-metal elements can stand in for the magnetic property of iron or the conducting capacity of copper.

Metal Ore Formation

NATURE LENT a hand in another crucial way in the development of metallurgy, in the ready availability of rocks and sediments rich in particular metals. Without these, the smelting of metals from the crustal rocks would never have occurred. And these metal ore deposits depend in turn on the recycling of the crustal rocks via plate tectonics. As the authors of *Planetary Crusts* explain, plate tectonics builds continents and forms ore deposits crucial to advanced civilization, thus "enabling this discussion to take place." From this the authors infer that any habitable and inhabited planet that develops high technology likely will require "plate tectonics and the evolution of continents" for concentrating "these elements into ore deposits."[34]

Plate tectonics is a huge and fascinating topic in itself, but any detailed consideration is beyond the scope of this chapter. Suffice here to say that the mechanism is enabled by another ensemble of fitness in nature. Marcia Bjornerud comments:

> Convection overturn occurs in the earth's interior only because of a fortuitous combination of physical variables. If rocks were better conductors of heat, such stirring would never take place, because the necessary temperature contrasts would be subdued. If rocks didn't expand significantly when heated there would be no density instability to drive convection. If the viscosity of Earth's mantle rocks were much higher, the whole system would grind to a halt. Finally if the planet had a smaller inventory of radioactive elements, or if these had much shorter half-lives, the bulb in the planetary lava lamp would have burned out long ago.[35]

And as we saw in Chapter 4, our planet's precious nitrogen, which acts as a fire retardant and makes safe fire possible, may also have been released from the crust as a result of active plate tectonics.[36]

Before Prometheus

THE MASTERY of fire, the development of metallurgy, and the technical advances which flowed from exploiting the properties of metals were only possible because of a vast ensemble of highly fortuitous prior elements of fitness in nature. While controlling fire, building kilns, and manufacturing charcoal were human discoveries, something else was needed, including wood fit for campfires and kilns and the fact that a readily obtained wood product, charcoal, possesses a combination of properties ideal for iron smelting and extraction. These were part of nature's regime long before humans. And it was not human genius which determined the properties of the carbon and oxygen atoms and their relative non-reactivity at ambient temperatures, which in turn rendered wood and vegetation, and indeed all organic carbon compounds, safe in the presence of high levels of atmospheric oxygen at ambient temperatures. Nor was it human genius that conferred on metals their uncanny utility for serving as the basis for a universe of useful tools, or their capacity to conduct electricity. Nor was it human genius that made available the rich ore deposits which our ancestors first mined and used to extract metals from the rocks.

The Twenty-First Century

WHAT MAKES the extraordinary fitness of nature for technology even more striking is that in addition to its fitness for safe fire and the use of wood and charcoal in the smelting of metals from their ores in ancient times, nature appears also extraordinarily fit for some of the most advanced technologies of this age.

Consider semiconductors. These are used in a host of electronic devices. But their invention was only possible because nature provided the right semiconducting materials for their construction, such as silicon crystals and several other materials, including the element germanium.

The rare earth elements (REE) also provide some extraordinary examples of the fitness of individual metals for specific twenty-first century ends. These unfamiliar metals, with exotic names such as Lanthanum, Yttrium, Neodymium, Dysprosium, Cerium, and Europium, have properties uniquely fit for many highly specific applications. A sheet from the US Geological survey notes that "the diverse nuclear, metallurgical, chemical, catalytic, electrical, magnetic, and optical properties of the REE have led to an ever increasing variety of applications. These uses range from mundane (lighter flints, glass polishing) to high-tech (phosphors, lasers, magnets, batteries, magnetic refrigeration) to futuristic (high-temperature superconductivity, safe storage and transport of hydrogen for a post-hydrocarbon economy)." The authors of the article go on to cite several specific examples of unique applications of the rare earth elements:

> Color cathode-ray tubes and liquid-crystal displays used in computer monitors and televisions employ europium as the red phosphor; no substitute is known.... Fiber-optic cables can transmit signals over long distances because they incorporate periodically spaced lengths of erbium-doped fiber that function as laser amplifiers. Er is used in these laser repeaters, despite its high cost (~$700/kg), because it alone possesses the required optical properties....
>
> Cerium, the most abundant and least expensive REE, has dozens of applications, some highly specific. For example, Ce oxide is uniquely suited as a polishing agent for glass....
>
> Permanent magnet technology has been revolutionized by alloys containing Nd, Sm, Gd, Dy, or Pr. Small, lightweight, high-strength REE magnets have allowed miniaturization of numerous electrical and electronic components used in appliances, audio and video equipment, computers, automobiles, communications systems, and military gear. Many recent technological innovations already taken for granted (for example, miniaturized multi-gigabyte portable disk drives and DVD drives) would not be possible without REE magnets....
>
> Several REE are essential constituents of both petroleum fluid cracking catalysts and automotive pollution-control catalytic converters.... A newly developed alloy, $Gd_5(Si_2Ge_2)$, with a "giant magneto-

caloric effect" near room temperature reportedly will allow magnetic refrigeration to become competitive with conventional gas-compression refrigeration."[37]

Another example of the fitness of nature for our technological civilization is witnessed in the great utility of so many regions of the electromagnetic spectrum outside the biological useful regions of the visual and near infrared (IR). As I wrote in *Children of Light*:

> Microwaves are used in every mobile phone and many other Wi-Fi devices. Similarly radio frequencies have been used for communication for more than 100 years, as through wireless and TV. X-ray crystallography was used to discover the structure of proteins and nucleic acids.... And of course X-rays, UV radiation, microwaves, and radio waves have been used along with light astronomy to vastly increase our understanding of cosmic and stellar evolution and of the overall structure of the cosmos.[38]

And so it turns out that nature is not just fit for the initial development of technology, which largely depended on the properties of the carbon, oxygen, and metal atoms and their ores. She is also specially fit for some of the most crucial technological advances upon which our twenty-first civilization now depends, some of which may well make possible the move to a new post-carbon electronic civilization over the next few decades.

Uniquely Fit

OUR BECOMING a fire maker, a tool maker, and eventually a voyager into space; our creating the *Curiosity* rover, bee-sized drones, and a laser that can illuminate the moon; all this has not been due to human genius alone but also to the fortuitous conditions on our planetary home as well as the fortuitous properties of so many of the elements of the periodic table and their molecular combinations. A unique fitness built into nature paved the way for our long journey of discovery and technological innovation.

Moreover, this path of discovery—from the stone age via fire to kilns, to ceramics, to metallurgy, to charcoal and kilns hot enough to smelt iron, and on via the industrial revolution to the technology of the twenty-first century—would appear to be unique, the only possible path given the laws of nature and the properties of matter as revealed by modern science. And what cinches this conclusion for me is the complete absence in the literature, to my knowledge, of any seriously developed hypothetical alternative route—dispensing, for example, with fire and metallurgy—that might have been taken by any type of intelligent life to arrive at an advanced civilization remotely on par with our own.

And as we shall see in the next chapter, that only biological beings similar to modern humans—possessed of our android design and size on a planet similar to Earth—could ever have exploited the wonderful fitness of nature for fire and metallurgy suggests that if there are, out there in the great dark, any extraterrestrial civilizations possessed of advanced technology, the denizens of those alien worlds will not only resemble ourselves, but their journey—while undoubtedly different in many fascinating details—will also in the main very much resemble our own.

11. THE FIRE MAKERS

But it will be asked for whose sake so vast a work was
carried out. Was it for the sake of trees and herbs, which
though without sensation are nevertheless sustained by
Nature? No, that at any rate is absurd. Was it for the
sake of animals? It is equally improbable that the gods
went to such pains for beings that are dumb and without
understanding. For whose sake, then, would one say that
the universe was formed? For the sake, undoubtedly, of
those animate beings that exercise reason.

—Marcus Cicero, *On the Nature of the Gods* (45 BC)[1]

Fire on Earth is a pervasive feature of the living world.
Life created the oxygen that combustion requires, and
provides the hydrocarbon fuels that feed it.... Fire takes
apart what photosynthesis has put together; its chemistry
is a *bio*-chemistry. Fire is not something extraneous to life
to which organisms must adapt, it is something that has
emerged out of the nature of life on Earth.

—Stephen Pyne, *The Ecology of Fire*[2]

IN THE PREVIOUS CHAPTER WE SAW THAT THE PATH FOLLOWED BY
our ancestors from stone tools to our current technological civilization was, in broad outline, the only possible path, and it was a path only
possible because of yet another stunning ensemble of prior environmental fitness in nature. In this chapter we will see that not only was the
"chosen path" unique, but that only a special type of unique being very

close to our own biological design could have taken the first and vital step to technological enlightenment, fire-making.

From first principles, a creature capable of creating and controlling fire must be an aerobic terrestrial air-breathing species, living in an atmosphere enriched in oxygen, supportive of both respiration and combustion. This fire maker must have something like human intelligence to accomplish the task, and while it is true that other species—e.g., dolphins, parrots, seals, apes, and ravens—possess intelligence and remarkable problem-solving abilities,[3] as far as is known, no other organism comes close to the intelligence of humans.

The species in question also needs to be mobile and possess high acuity vision in order to be able to create and master fire, and follow the subsequent route via metallurgy to an advanced technology.

Being a social species possessed of language would also have been essential for the peripheral tasks associated with the regular making and controlling of fire among small tribal groups, including the hewing and collecting of the necessary wooden fuels to initiate and sustain the fire. While many other species are social, none possesses a communication system remotely as competent as human language for transmitting information, including abstract concepts.[4]

In addition to being terrestrial, air breathing, sighted, mobile, intelligent, social, and possessed of language, a fire-maker also needs the right anatomy. And in keeping with the anthropocentric claim, only humankind of all the creatures on Earth is properly endowed with the right build to make and control fire. Neither a giraffe, nor an elephant, nor a parrot, nor a cat, nor a chimp nor any other terrestrial organism has the right anatomy to master fire, none apart from humans.

Hands and Arms

As ANY boy scout or girl scout knows, starting a fire by rubbing two sticks together requires considerable manual dexterity and persistence. And as Tom Hanks's character found out in the film *Castaway*, even with the superb manipulative abilities of the human hand, it is difficult

Figure 11.1. Two men on the Kalahari starting a fire with a fire drill.

to start a fire using traditional frictional methods such as rubbing two pieces of wood together. However, with practice and using various simple devices such as a fire drill, most humans can master the skills necessary to initiate a fire.

No other species on Earth possesses an organ remotely as capable as the human hand for initiating and maintaining a fire and for intelligent exploration and manipulation of the natural world. One of the earliest and still most fascinating discussions of this adaptive marvel was given by the first century Greek physician Galen:

> To man the only animal that partakes in the Divine intelligence, the Creator has given in lieu of every other natural weapon or organ of defence, that instrument the hand: applicable to every art and occasion.... Let us then scrutinize this member of our body; and enquire, not simply whether it be in itself useful for all the purposes of life, and adapted to an animal endowed with the highest intelligence; but whether its entire structure be not such, that it could not be improved by any conceivable alteration.[5]

Galen then proceeds for several pages to detail the genius of the human hand. At one point he declares, "Whoever admires not the skill and contrivance of Nature, must either be deficient in intellect, or must have some private motive, which withholds him from expressing his admiration. He must be deficient in intellect, if he do not perceive that the human hand possesses all those qualifications which it is desirable it should possess; or if he think that it might have had a form and construction preferable to that which it has."[6]

The perfection of the hand was a popular topic among nineteenth-century natural theologians such as Oxford Professor of Medicine John Kidd, who quoted liberally from Galen. Charles Bell, in his Bridgewater Treatises, also waxed lyrical about the hand, while also citing one of the ancient Greek thinkers: "Seeing the perfection of the hand, we can hardly be surprised that some philosophers should have entertained the opinion with Anaxagoras, that the superiority of man is owing to his hand.... it is in the human hand that we have the consummation of all perfection as an instrument."[7]

Only the great apes, our cousins, come close. Yet the hand of the chimp and gorilla, although possessing a partially opposable thumb, is far less adapted to fine motor movement and control than is the human hand with its fully opposable thumb. Although some chimps exhibit a remarkable manual dexterity for certain tasks,[8] none can match the manual dexterity of the human hand.[9] And this is obvious on watching chimps at a "tea party" at the zoo. A dining task we hardly think about proves comically challenging for them due to their limited manual dexterity.

Science journalist Christopher Joyce reviewed the unique capacities of the human hand in a 2010 article at NPR, drawing on insights from two of his interviewees, anthropologists Erin Marie William and Caley Orr:

> Now, apes make tools. Scientists have trained a bonobo, called Kanzi, to do that. But Kanzi's not much good at it.
>
> "He just can't get the motions down," Williams says. That's because he can't grip the stones, his thumbs aren't long enough and his

fingers are too long and he's clumsy. He can't move his wrists—he can't extend his wrist and get this important "snap." He makes a mess.

... Anthropologist Caley Orr... has laid out the skeletal hands of three apes and a human. The apes' hands are enormous—the orang-utan's is like a catcher's mitt. But their thumbs are tiny and splayed out to the side; the fingers are long and curved. They look powerful, but Orr says the strength runs vertically, from the wrist up through the fingers. That's good for hanging on tree limbs, but not for much else.

.... The human hand is smaller, and it works differently. Orr hands me a two-foot-long club to illustrate.

"Here, try to hold this without using your little finger, and just using those other digits," he says. That's the way an ape might hold it. I make to swing it but realize it will fly out of my hand if I do.

The strength in my hand extends across my palm. My thumb is stronger, and so is my pinkie. I can wrap that thumb over my other fingers and then secure the grip at the bottom with my pinkie. An ape can't manage that very well.

And my opposed thumb and wider fingertips also mean I can grip a round stone—like a hammerstone—with more control than an ape can.

I have the hand of the ultimate toolmaker.[10]

The hand's attachment to the end of a highly mobile appendage about two-and-a-half feet long—the human arm—further contributes to its universal utility and enables the hand to manipulate objects some distance from the body, no small advantage when manipulating fire. Moreover, because of the hand's positioning at the end of the arm, its manipulative activities can be readily observed by our eyes, which are placed forward-facing on the head so that the activities of the hand can be readily observed.

Other primates have arms of sufficient length for manipulating fire, but their arms and hands have another role that handicaps them for the work of adroitly manipulating tools. All the great apes are basically quadrupeds, or more precisely, knuckle-walkers as defined by Richard Owen in the nineteenth century.[11] Our upright bipedal gait and android design frees the human arms and hands from the ambulatory function,

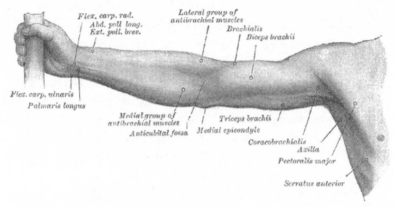

Figure 11.2. The human arm and hand.

allowing them to take on adaptations for delicate activities requiring fine motor control. Among primates, a habitual bipedal posture is only present in humans and in a handful of fossil hominin species. This bipedal posture enabled our arms and hands to acquire their unique manipulative functions, which in turn enabled our ancestors to initiate and control fire, and carry out the peripheral activities associated with fire-making, such as hewing and collecting wood, as well as the activities associated with the subsequent development of metallurgy such as mining for ores and building kilns, not to mention the ability to construct a great diversity of tools and instruments, the use of which has been crucial to the development of technology and advances in scientific knowledge.

The Right Size

POSSESSING SUCH a superb manipulative organ as the hand and an upright bipedal android form that frees the arms and hands for purely manipulative tasks are not in themselves sufficient. They would be of no avail unless we were the right size. Only an android organism of approximately our dimensions can readily make and control fire. To illustrate, let's begin very small and then work up. An android organism the size of an ant—like Ant Man in the Marvel comics—would be far too small because the heat would kill him long before he was even several body lengths from the flames.[12] As Hu Berry points out, "Ants cannot use fire, for the simple reason that the smallest, stable fire must be much larger than an ant. Ants cannot therefore carry fuel close enough to a fire to maintain it."[13]

Even diminutive humans two feet tall, possessed of our android design and all our unique anatomical adaptions, would face enormous difficulties in attempting to manipulate fire. Although the recently discovered species of diminutive humans *Homo floresiensis*[14] did use fire, it seems likely that a species any smaller than their reported height of 3.5 feet would have considerable difficulty.

There is another consequence of being smaller. In a fascinating article in the *American Scientist* entitled "The Size of Man," author F. W. Went points out that organisms much smaller than humans lack the ability to generate the necessary kinetic forces to procure the essential raw materials for fire making and metallurgy. This is because the kinetic energy generated when a mass moves a particular distance (the head of a hammer striking a nail, an axe hitting a tree trunk, a pick hitting an ore deposit) varies as the length traversed raised to the fourth power (kinetic energy = KL^4).[15] And this means, as Went explains, that "if we assume that a spear or a club of a size proportionate to body size is handled by a 7 ft giant, its impact would be 4 times greater than when handled by an ordinary 5 ft 8 in. man... But compared with an ordinary man, the blow of a 3-year old child or a 3 ft creature in general could only produce one twenty-fifth of the energy, utterly insufficient to kill prey or hunt larger animals." Consequently, "a 3-ft man could neither cut lumber nor excavate a mine in solid rock."[16]

Stephen Jay Gould also addressed the issue of bodily size, noting that kinetic energy increases with length raised to several orders of magnitude. He goes on to confess "a special sympathy for the poor dwarfs who suffer under the whip of cruel Alberich in Wagner's *Das Rheingold*. At their diminutive size, they haven't a chance of extracting, with mining picks, the precious minerals that Alberich demands."[17]

So our size is right for approaching a fire, for generating the necessary kinetic forces to hew the wood needed to fuel high-temperature fires, and for mining metal ores from rocks. We could be neither fire makers nor metallurgists if we were significantly smaller.

On the other hand, it is fortunate that the ability to hew wood and mine for ores does not necessitate kinetic forces much greater than those that can be generated by organisms of our dimensions. While significantly larger android beings could exert greater kinetic forces, the design of a bipedal primate of, say, twice our height would be severely constrained by kinetic and gravitational forces and be structurally problematic.

Why? For one, and as discussed in Chapter 9, mass (and weight) increases by the length cubed (L^3) while the strength of bone and the power of muscles increases only by the length squared (L^2). J. B. S. Haldane alluded to this with characteristic lucidity in his essay "On Being the Right Size":

> Consider a giant man sixty feet high—about the height of Giant Pope and Giant Pagan in the illustrated *Pilgrim's Progress* of my childhood. These monsters were not only ten times as high as Christian, but ten times as wide and ten times as thick, so that their total weight was a thousand times his, or about eighty to ninety tons. Unfortunately, the cross-sections of their bones were only a hundred times those of Christian, so that every square inch of giant bone had to support ten times the weight borne by a square inch of human bone. As the human thigh-bone breaks under about ten times the human weight, Pope and Pagan would have broken their thighs every time they took a step. This was doubtless why they were sitting down in the picture I remember.[18]

There is yet another kinetic constraint on being too big, arising again from the fact that kinetic energy is proportional to L^4. Went explains:

> The numerical values of kinetic energy actually give us a good clue as to the optimal size of man. A 2 m tall man, when tripping, will have a kinetic energy upon hitting the ground 20–100 times greater than a small child who learns to walk. This explains why it is safe for a child to learn to walk; whereas adults occasionally break a bone when tripping,

children never do. If a man were twice as tall as he is now, his kinetic energy in falling would be so great... [16 times more than at normal size] that it would not be safe for him to walk upright.... The larger mammals can become taller, because they are more stable on their four legs. Yet, they break bones more easily when they fall.[19]

As Steven Vogel puts it in *Comparative Biomechanics*, "Tripping is a potential danger to cows, horses, and the like... we run a similar risk even at a lower body mass; the upright posture of humans gives us an unusually great height relative to our mass."[20] He summarizes the larger point by recourse to the old saying, "The bigger they are, the harder they fall."[21]

Consequently tripping and falling would be a catastrophe for Pope or Pagan as their massive heads (1,000 times the volume of a human head) would hit the ground with such force that the skull would be fragmented and the brains disintegrate.

The Right Size Planet

There is no escape from the above problem by envisaging a giant like Pope or Pagan on a smaller planet where gravity and kinetic forces presented less of a challenge. Planets significantly smaller than the Earth, where gravitational and kinetic constraints would be less, tend to lose their atmospheres and, with them, their precious oxygen, so they could not host large active advanced aerobic organisms, which depend on oxygen extracted directly from an atmosphere to satisfy their metabolic needs. And even if they could somehow exist under such oxygen-starved conditions, there could be no fire-making and, therefore, no path to high technology.

On the other hand, planets appreciably larger than Earth would possess gravity that would exacerbate the dangers of tripping. Also, on such planets the weight of any animal's body would be increased, imposing severe constraints on the capacity of the muscles to empower movement and support an upright bipedal stance. An ant would be fine on a planet where gravity was several times stronger, but as we have seen above, fire-making necessitates an organism considerably larger than an ant. Worse, such outsized planets retain their primeval mix of hydrogen and helium and end up as gas giants like Jupiter and Saturn.

Earth-sized planets are neither too big nor too small. Only such planets have the right oxygen-rich atmosphere to support fire-making and human respiration, and the right gravity (at or near 1 g) to allow for upright android beings of our dimensions. Additionally, the strength of muscles is just sufficient to endow organisms of our size and weight with mobility and the ability to stand erect at 1 g.

So what we have is a remarkable anthropocentric coincidence: the gravity of planets capable of retaining atmospheric oxygen—needed for high-metabolism active terrestrial aerobes with the intelligence needed for mastering fire—is also just the right gravity to enable the functioning of beings of our size and dimension, beings capable of making and controlling fire.

The Right Inertia

INERTIA IS the name we give to the property of things to resist a change in their velocity. An undisturbed body remains at rest and requires the exertion of a force to impart motion to it. A moving car requires force to slow it down or to make it change direction. Like gravity, inertial forces are related to mass. It requires more force to make a large object move or change direction than for a small one. If inertia were significantly less, the wind could set a boulder rolling. In such a world we would be continually bombarded by all manner of objects. If inertia were far greater, one would have difficulty even moving a hand, and once in motion, control of its direction and speed would be equally difficult. The inertia of matter must be fairly close to what it is for creatures of our size to function in an environment similar to Earth's.

Also, lighting a fire in a counterfactual universe where inertial forces were significantly less would pose profound dangers. The logs and burning embers would be forever drifting off every time a gust of wind passed through, thereby igniting nearby vegetation. Safe, controllable fire would be impossible. On the other hand, if inertial forces were greater, collecting fuel for the fire might prove an insurmountable task, with even small logs proving very nearly impossible to move. That the inertial force is set where it is allows us to exist in a habitable environment, and to make and feed controllable fires.

Where the inertial force come from, and what causes it to take the precise strength it does, remain unsolved questions in physics,[22] but one thing is clear: inertia's particular strength stands as another example of prior fitness, not just for life but for creatures such as ourselves.

In Sum

WITH THE evidence of our unique fitness for fire-making, which allowed humankind to take the first and fateful step along the unique path to an advanced technological civilization, the final piece of my argument built up over the previous chapters is complete. Nature is uniquely fit not just for our biological being, for our aerobic terrestrial existence and for our size and body plan, but also for our unique capacity for fire-making and for following the singular path through metallurgy to an advanced technology and a profound understanding of nature.

That claim can only be challenged by showing that, given the laws of nature and the structure of the cosmos, there is the possibility of fundamentally different types of biological life than ones based on carbon and water, the possibility of fundamentally different types of intelligent beings capable of making and controlling fire, and fundamentally different routes to an advanced technology and deep knowledge of the world, routes that do not pass through fire-making and metallurgy. However, no credible alternatives of this sort have ever been proposed. No single book or paper exists which provides a well-worked-out blueprint for a cell radically different from the canonical carbon-based cell, the basic building block of all life on Earth. No single paper or book exists which describes within the domain of carbon-based life an alternative biological design for an advanced organism comparable to modern humans possessed of a high metabolic rate and high intelligence. Nor has an alternative design ever been proposed in any detail for an advanced intelligent aerobic organism capable of making and controlling fire. Nor is there a single paper which describes a substantially different route to an advanced technology and ultimately scientific knowledge. One can imagine variations on the humanoid theme, of course, some more realistic than others. But to the degree that they are credible, they will be variations very close to the human form.

Even though many mysteries remain, we can now, in these first decades of the twenty-first century, at last answer with confidence Thomas Huxley's question of questions as to "the place which mankind occupies in nature and of his relations to the universe of things."[23] As matters stand, the evidence increasingly points to a natural order uniquely fit for life on Earth and for beings of a biology close to that of humans, a view which does not prove but is entirely consistent with the traditional Judeo-Christian framework. Humankind's exile from nature, which to many people's minds commenced in the sixteenth century with the demise of the geocentric model of the universe, itself appears to be meeting its demise as evidence mounts that the logos, the underlying rationality of all things, is indeed "manifest in human flesh."

The wheel has turned. Scientific advances beginning with the flowering of chemistry in the nineteenth century and continuing at ever-increasing pace through the twentieth century and now into the twenty-first have vindicated the ancient covenant and revealed humanity to be as the medieval scholars believed, reflective in the depths of his natural being of all facets of the greater macrocosm of which he is an integral part.[24] And in one of history's supreme ironies, it is now the denial of humankind's special place in nature, the foundational denial of the current secular culture and *Zeitgeist*, which grows increasingly outdated and devoid of empirical support.

12. The End of the Matter

The ancient opinion that man was microcosmus, an abstract or model of the world, hath been fantastically strained by Paracelsus and the alchemists, as if there were to be found in man's body certain correspondences and parallels, which should have respect to all varieties of things, as stars, planets, minerals, which are extant in the great world.

—Francis Bacon, *The Advancement of Learning* (1605)[1]

From 1953 onward, Willy Fowler and I have always been intrigued by the remarkable relation of the 7.65 Mev energy level in the nucleus of C12 to the 7.12 Mev level in O16. If you wanted to produce carbon and oxygen in roughly equal quantities by stellar nucleosynthesis, these are the two levels you would have to fix, and your fixing would have to be just where these levels are actually found to be.... A common sense interpretation of the facts suggests that a superintellect has monkeyed with physics, as well as with chemistry and biology, and that there are no blind forces worth speaking about in nature.

—Fred Hoyle, *Engineering and Science*[2]

I CONFESS THAT THE CLAIM DEFENDED HERE—THE ANTHROPOCEN-tric claim that the cosmos is uniquely fit for the existence of beings of our physiological design and biology—will strike many of my contemporaries as outrageous. The claim is indeed extraordinary, and as Carl Sagan has said, "Extraordinary claims require extraordinary evidence."[3]

Showing that there is indeed extraordinary evidence in support of the anthropocentric conception of nature has been the aim of this and the previous monographs in this Privileged Species series. No matter how unfashionable the idea, no matter how extraordinary the claim, the facts speak for themselves.

The human person as revealed by modern science is no contingent assemblage of elements, an irrelevant afterthought of cosmic evolution. Rather, our destiny was inscribed in the light of stars and the properties of atoms since the beginning. Now we know that all nature sings the song of man. Our seeming exile from nature is over. We now know what the medieval scholars only believed, that the underlying rationality of nature is indeed "manifest in human flesh." And with this revelation the post-Copernican delusion of humankind's irrelevance on the cosmic stage has been revoked.

As things now stand, the current evidence points irresistibly to a natural order uniquely fine tuned for life on Earth and for beings of our biological design. We are indeed favored "light eaters" and "fire makers" in the grand cosmic scheme. Fourteen billion years before our origin in the Pleistocene, our biological design was prefigured in the order of things at the moment of creation.

Oxygen Redux

PERHAPS NO scientific discovery in the century since Lawrence Henderson's seminal work, *The Fitness of the Environment*, provides such compelling evidence of a special environmental fitness in the order of the world for advanced beings of our biology than the story of oxygen. In the tale of oxygen, science reveals a vast suite of diverse elements of fitness in nature that enable the existence of air-breathing terrestrial aerobes like ourselves as well as our ability to make and master fire, an ensemble unimagined little more than a century ago.

As we have seen, this stunning chain of unique elements of fitness includes the fitness of sunlight to provide the right radiation (with the right energy levels) for photochemistry and to warm our planet, as well as the fitness of our atmospheric gases, which let through just the right radiation for photochemistry while absorbing a significant fraction of the infrared radiation, which raises the temperature of Earth's surface well above freezing. It also includes oxygen's reluctance to react, which in conjunction with nitrogen's fire-retarding property, allows us to coexist with an atmosphere enriched in oxygen without undergoing spontaneous combustion, and renders fire safe and available for technological exploitation.

Then there is the low density and viscosity of air. This ensures that breathing is not too demanding, and commensurate with the body's energy budget. There is also the high diffusion rate of gases, critical in enabling the uptake across the 130 square meters of alveolar membrane of the 250 ml of pure oxygen we need every minute to supply our energy needs.

The transport of the oxygen to the tissues from the lungs depends on a circulatory system which in turn depends on the existence of a liquid, water, with just the right properties to serve as the medium of the blood—water being an ideal solvent and having the right viscosity, density, and so forth. The transport of oxygen in the blood also depends on a carrier device with the right properties to gently bind and dissociate with the oxygen molecule, and again nature obliges, because the transition metals like iron have just the right properties to help carry out this vital task. No alternatives that might replace the transition metals are available.

And consider the end point of the oxygen cascade. Here the oxygen reaches the mitochondria and reacts with the body's reduced carbon fuels. This generates carbon dioxide (CO_2), a substance which, being a soluble gas at ambient temperatures, has just the right properties for excretion from the body via the lungs and just the right properties to react with water to produce bicarbonate (HCO_3), which in turn is just right for carrying most of the CO_2 to the lungs for excretion, and for buffering an air-breathing organism against changes in the acidity of the blood.

In short, science has revealed that our existence as energy-demanding active air-breathing terrestrial organisms critically depends on a wildly improbable ensemble of natural environmental fitness comprising various chemical and physical laws as well as the properties of specific molecules such as oxygen and CO_2 and specific elements such as the transition metals, properties that must be almost exactly as they are. And most of this prior ensemble of fitness is of little or no relevance to the great mass of carbon-based life forms on Earth, including the anaerobic denizens in the crustal rocks and the menagerie of water-breathing aquatic organisms. It is a fitness for land-based oxygen-hungry aerobes like ourselves.

It is now widely acknowledged that carbon-based life embedded in a water matrix is the only type of life permitted by the laws of nature.[4] Given this and the necessity for oxidation to supply the energy needs of complex carbon-based life throughout the universe, and given the necessity to be land-based aerobes taking in oxygen directly from an atmosphere to derive the greatest amount of energy from oxidations and thereby fuel our large energy-hungry brains, it follows that all intelligent life anywhere in the universe will be oxygen-hungry complex beings of a biological design very close to our own, possessing a circulatory system like our own, possessing lungs, using transition metals to tame and utilize oxygen, and thriving in an atmosphere containing a pO_2 of about 160 mm Hg. They will strongly resemble *Homo sapiens*.

The only explanation that makes sense of this extraordinary ensemble of prior fitness for advanced aerobic terrestrial life is the traditional anthropocentric view of the cosmos as a uniquely fit whole for beings of our biology, that we do indeed occupy a special place in the natural order.

Water Redux

As FOR water, what explanation makes sense of its unique set of features that make possible the hydrological cycle? This cycle, recall, delivers water to the land, and depends on water's anomalous property of existing in three material forms in ambient conditions, in conjunction with its mobility, its powers as a solvent, and its high surface tension. These work together to weather and erode rocks, providing water enriched in minerals for life on the land, while at the same time creating the soils that serve as the ideal substrate for the growth of plants.

Water also possesses the highest specific heat of any common fluid and the highest latent heat of evaporation (cooling) of any molecular substance,[5] two thermal properties beautifully fit to enable the existence of warm-blooded terrestrial air-breathing organisms like ourselves. And adding wonder to wonder, the same fluid has just the right properties to form the medium of a circulatory system, an essential adaptation for all large multicellular life forms, and in the case of warm-blooded organisms like ourselves, convey the excess heat generated by our high metabolic rate to the skin where, in perfect concordance with the anthropocentric paradigm, water's outstanding property as an evaporative coolant draws the excess heat of metabolism out of the body.

The Right Proportions

AND WHAT other paradigm makes sense of the staggering improbability, given the great number of diverse physical constants involved, that the functional volumes of the various organ systems should be commensurate with the mammalian and human body plan, and so can be accommodated and function together in the same organic form? Or that high acuity vision is possible in an organ the size of the eye? Yet further miracles of fitness built into nature from the beginning.

Fire Makers

NATURE IS also fine tuned for our mastery and control of fire. As we saw in Chapter 4, this depended on an ensemble of prior fitness: the attenuation of the reactivity of oxygen at ambient temperatures, the existence of the diluent nitrogen in the atmosphere, and the fact that the same partial pressure of oxygen in the atmosphere supports both human respiration and combustion (fire). And as we saw in Chapter 10, there was an equally remarkable ensemble of fitness enabling humankind to follow what would appear to be a unique route from the stone age via fire to ceramics, metallurgy, charcoal-fueled kilns hot enough to smelt iron, and via the industrial revolution to our current technological civilization.

Prior Fitness: The Blind Spot

GIVEN SUCH extraordinary evidence for a very special environmental fitness for human biology and even for the path we took from the stone age to the twenty-first century, a reader might well question why, if the evidence is so abundant, does the mainstream in biology "look the other way." The reason for this blind spot among evolutionary biologists may be that, as Henderson stressed over a century ago, biologists since Darwin have focused almost entirely on the fitness (or adaptations) of organisms to the environment, and not on the prior environmental fitness that enables the actualization of the adaptations. As Henderson put it:

> But although Darwin's fitness involves that which fits and that which is fitted, or more correctly a reciprocal relationship, it has been the habit of biologists since Darwin to consider only the adaptations of the living organism to the environment. For them, in fact, the environment, in its past, present, and future, has been an independent variable, and it has not entered into any of the modern speculations to consider if by chance the material universe also may be subjected to laws which are in the largest sense important in organic evolution. Yet fitness there must be, in environment as well as in the organism. How, for example, could man adapt his civilization to water power if no water power existed within his reach?[6]

And how indeed could there be terrestrial life without the prior environmental fitness of the hydrological cycle? How could there be aerobic life without nature's prior fitness for photosynthesis and without the special attenuation in the reactivity of oxygen in ambient conditions? How could there be high acuity vision in an organism about our size unless the wavelength of light was almost exactly where it is? And how could there be a circulatory system without the prior properties of water? And how could there be warm-blooded terrestrial organisms without the thermal properties of water?

The omission of any acknowledgement of the vital significance of prior environmental fitness is one of the great blind spots of evolutionary biology. Amazingly, there is not a single mention in *The Origin of Species* of the crucial role of the fitness of the environment and of the laws of physics and chemistry for life. And the same mystifying absence of any acknowledgement of prior environmental fitness in enabling so many of the grand innovations and adaptations in the history of life is obvious on any reading of any of the major evolutionary texts since *The Origin*.[7]

In the Depths of Nature

IT IS one of the great ironies in the history of Western thought that the medieval anthropocentric paradigm, considered an archaic anachronism by most modern commentators, makes more sense of the twenty-first century scientific evidence than the nihilistic presumptions of such leading Darwinists as Richard Dawkins and Jacques Monod. According to historian of philosophy Frederick Copleston, for the fifteenth-century German theologian and philosopher Nicholas of Cusa, "Although each finite thing mirrors the whole universe, this is particularly true of man,"[8] and for the ninth-century Irish scholar John Scotus, "Man is the microcosm of creation, since he sums up in himself the material world and the spiritual world... he is in fact... the link between the material and spiritual, the visible and invisible creation."[9] For the medieval scholars humankind thus occupied a central place in the world created by God, a conception of nature far closer in essence to the scientific picture that has emerged since the mid-nineteenth century than Stephen Jay Gould's celebrated view (speaking sadly for a majority of present-day biologists) of *Homo sapiens* as "a tiny twig on an improbable branch of a contingent limb on a fortunate tree... a detail, not a purpose."[10]

And what was the source of the medieval anthropocentric view? It was not modern scientific knowledge. For while many of the schoolmen were brilliant philosophers, their knowledge of the natural world was very rudimentary. The geocentric model played a role, as did the writings of classical philosophers,[11] but the primary source of their conviction that the cosmos was arranged for humankind was, more than any factor, the Judeo-Christian scriptures.

The evidence presented in these pages builds what can rightly be described as an overwhelming case for the anthropocentric claim that the laws of nature are uniquely fit to provide the necessary environmental conditions to enable the existence of beings of our physiological and anatomical design, as well as enabling beings such as ourselves to follow the singular technological path from the Stone Age to our current advanced technological civilization and acquire a deep knowledge of the mysterious universe from which we emerged.

The discoveries reviewed in this book have revealed that there is, after all, a profound connection between man and cosmos, between biology and physics, one based not on the Judeo–Christian scriptures or classical philosophy but on evidence derived from advances in our scientific understanding of nature. The ensembles of prior environmental fitness for our biology described in the previous chapters provide the basis for a powerful and newly reinvigorated scientific reformulation of the traditional anthropocentric paradigm. And this has, as mentioned above and in previous chapters, the intriguing and purely secular implication that if there are other advanced intelligent organisms in the universe, and especially ones possessing advanced technology, they will inhabit a planet resembling Earth and closely resemble ourselves: terrestrial air-breathing carbon-based life forms, about our size, android in form, who will have reached an advanced technology via the same general route followed by our ancestors, via fire-making and metallurgy to a scientific revolution and on towards technologies that exploit the rare earth elements.

A skeptic might assert that things have to be fit for us or we would not be here to notice our good fortune. And of course this is true, but it is entirely trivial, and it misses the point of the core argument defended in this book. My argument is not merely that nature is fit for us (which it must be, of course), but that nature is *uniquely fit* for intelligent, technologically capable organisms very much like us, that we occupy a very special, even privileged, place in the order of things. That is the central claim of this book. And it is a claim which, as I have shown, is supported by a mountain of scientific evidence.

Humans are clearly no contingent cosmic afterthought. The exquisitely fine-tuned ensembles of environmental fitness described here, each enabling a vital aspect of our physiological design, amount to nothing less than a primal blueprint for our being written into the fabric of reality since the moment of creation, providing compelling evidence that we do indeed, after all, occupy a central place in the great cosmic drama of being.

This is the miracle of man. We are not positioned in the spatial center of the universe as was believed before Copernicus, but what we have found over the past two centuries confirms the deep intuition of the medieval Christian scholars who believed that "in the cognition of nature in all her depths, man finds himself."[12]

ENDNOTES

1. INTRODUCTION

1. Lawrence J. Henderson, *The Fitness of the Environment: An Enquiry into the Biological Signifi-cance of the Properties of Matter* (New York: McMillan, 1913), 312.

2. *The Miracle of Man* presents a good deal of fresh evidence not covered in any of my previous works, but in providing in this final book in the Privileged Species series a comprehensive overview of the most significant elements of environmental fitness for our unique biologi-cal design and technological journey, inevitably I revisit some of the evidence covered in my previous books and journal articles. I have provided a liberal sprinkling of signposts of this indebtedness, both in the main text and in the endnotes, for those interested in delving further.

3. It should be noted that the medieval period could also be characterized as theocentric, and in the medieval and Renaissance periods the relationship between anthropocentrism and geocentrism is more complex than is generally suggested in popular accounts. Earth, for example, was viewed by medieval scholars not so much as the exalted center of the universe but as the bottom of the universe, far below the Empyrean where God dwells. In the Re-naissance Copernicus, at least, argued against a stand-together/fall-together coupling of geocentrism and anthropocentrism, insisting he was convinced that "the motions of the mechanism of the universe" that he did so much to illuminate had been "established for us [humans] by the best and most systematic craftsman of all." (Nicolaus Copernicus [1543], *On the Revolutions of the Heavenly Spheres*, trans. Alistair Matheson Duncan (United Kingdom: David & Charles, 1976), 25.) Other early champions of the heliocentric model also argued that the displacement of the geocentric model did not demean Earth and hu-mankind. For example, Galileo wrote, "Many arguments will be provided to demonstrate a very strong reflection of the sun's light from the Earth—this for the benefit of those who assert, principally on the grounds that it has neither motion nor light, that the Earth must be excluded from the dance of the stars. For I will prove that the Earth does have motion, that it surpasses the moon in brightness, and that it is not the sump where the universe's filth and ephemera collect." (Galileo Galilei [1610], *Siderus Nuncius*, quoted in Dennis Danielson, ed., *The Book of the Cosmos: Imagining the Universe from Heraclitus to Hawking* (New York: Basic Books, 2000), 150.) In the same year, Kepler also pressed the case that the heliocentric model was well worthy of man and man of it. As he commented, "In the interests of that contemplation for which man was created, and adorned and equipped with eyes, he could not remain at rest in the center," as in the case of the geocentric model. "On the contrary, he must make an annual journey on this boat, which is our Earth, to perform his observations." After all, only by moving around the sun could the human astronomer on planet Earth discover so much about the heavens. (Johannes Kepler [1610], *Kepler's Conver-sation with Galileo's Sidereal Messenger*, trans. Edward Rosen (Johnson Reprint Corporation,

1965), 45. At the same time it is certainly possible to see in these statements something like men standing at the bar, arguing against a popular accusation, namely that the heliocentric model they were propounding did indeed diminish man and Earth. Galileo himself points out, in *Siderus Nuncius*, that the moon does not make Earth special, as Jupiter has four moons. He makes plain that even our solar system is not unique, as it has a small analogue in the Jovian system. And the second half of *Nuncius* is about the vast number of additional stars revealed through Galileo's telescope, the first step towards what was to be a growing realization that the number of other "suns" in the universe is far greater than previously imagined. And as Dennis Danielson notes, for more than a few the exhilaration of the heliocentric discovery "translated into bewilderment. One thinks of John Donne's oft-quoted lament, 'Tis all in pieces, all coherence gone;' or Pascal's 'The eternal silence of these infinite spaces frightens me;' or Robert Burton's humorous but frustrated roundup of the cosmologists of his day: 'the world is tossed in a blanket amongst them, they hoist the Earth up and down like a ball.'" (Dennis Danielson, "The Great Copernican Cliché," *American Journal of Physics* 69, no. 10 (October 2001), 1033, DOI:10.1119/1.1379734. Internal references removed.) Danielson adds that by the middle of the seventeenth century, writers can be found who explicitly associated geocentrism with human self-importance and construed geocentrism's demise as a grave blow to that self-importance: "Among these are Cyrano de Bergerac, who protests 'the insufferable pride of humans,' and Thomas Burnet, who as it were retaliates by referring to our Earth as an 'obscure and sordid particle.' But it is the great French popularizer of Copernicanism Bernard le Bouvier de Fontenelle who most powerfully asserts the negative axiological implications of the new cosmology: In his famous *Entretiens sur la Pluralite´ des Mondes* [*Conversations on the Plurality of Worlds*], the lady in the dialogue, upon hearing about the heliocentric model, declares that Copernicus, had he been able, would have deprived Earth of the moon just as he has deprived it of all the other planets, for she perceives, she says, that he 'had no great kindness for the Earth.' Yet Fontenelle's own character replies to the contrary by praising Copernicus: 'I am extremely pleased with him... for having humbled the vanity of mankind, who had usurped the first and best situation in the universe.'" Danielson continues: "This interpretation of Copernicanism became the standard and apparently unquestioned version of the Enlightenment, as magisterially summarized by Goethe: 'Perhaps no discovery or opinion ever produced a greater effect on the human spirit than did the teaching of Copernicus. No sooner was the Earth recognized as being round and self-contained, than it was obliged to relinquish the colossal privilege of being the center of the universe.' And from Goethe and the Enlightenment to the present there has been, in more senses than one, almost no looking back." (Danielson, "The Great Copernican Cliché," 1034. Internal references removed.)

4. C. M. Cipolla and A. Grafton, *Clocks and Culture, 1300–1700* (New York: Norton, 2003), 31–44.

5. In *Evolution: Still a Theory in Crisis* I wrote, "Indeed, from Aristotle down, throughout the medieval period, right up to the seventeenth century, life was always conceived to be an integral part of nature, and its constituent forms—substantial forms—basic components of the world-order. Aristotle conceived of these as active agents in nature, molding the forms of organisms and, through their collective activities, the overall pattern of life on earth. As Jonathan Lear comments regarding Aristotle's conception of forms: 'Since the seventeenth century Western science has moved steadily away from conceiving forms as part of the basic fabric of the universe... In Aristotle's world, forms... occupy a fundamental ontological position: They are among the basic things that are.' After being in the cold for most of the past 150 years, overshadowed by the 'cult of the artifact,' the traditional notion that life is an integral part of the natural order has found renewed support in the revelation of twentieth-

century physics and cosmology that the laws of nature are uniquely fine tuned to a remarkable degree to generate environmental conditions ideal for life as it exists on earth." (Seattle, WA: Discovery Institute Press, 2016), 247–248. The Jonathan Lear quotation is from his book *Aristotle: The Desire to Understand* (New York: Cambridge University Press, 1988), 20.

6. Aron Gurevich, *Categories of Medieval Culture* (London: Routledge and Kegan Paul, 1985), 57–61.

7. S. J. Dick, *Many Worlds* (Philadelphia: Templeton Foundation Press, 2002), xii.

8. Alexandre Koyré, *From the Closed World to the Infinite Universe* [1957] (New York: Harper & Brothers, 1958), 4.

9. John Donne, "An Anatomy of the World" [1611], https://www.poetryfoundation.org/poems/44092/an-anatomy-of-the-world.

10. William Harvey, *Exercitatio Anatomica de Motu Cordis et Sanguinis in Animalibus* [1628]. For a facsimile of the original Latin publication, along with an English translation, see *Exercitatio Anatomica de Motu Cordis et Sanguinis in Animalibus* (Springfield, IL: Charles C. Thomas, 1928), https://archive.org/details/exercitatioanato00harv.

11. Robert Hooke, *Micrographia* (London: Jo. Martyn and Ja. Allestry, 1665), https://archive.org/details/mobot31753000817897/page/n1.

12. Koyré, *From the Closed World to the Infinite Universe*, v.

13. Jacques Monod, *Chance and Necessity* (London: Collins, 1972), 49.

14. Stephen Jay Gould, *Wonderful Life: The Burgess Shale and the Nature of History* (New York: Norton, 1990), 319.

15. Gould, *Wonderful Life*, 291.

16. Carl Sagan, *Cosmos* (New York: Ballantine Books, 2013), 17.

17. William Whewell, Bridgewater Treatise no. 3, *Astronomy and General Physics Considered with Reference to Natural Theology* (London: William Pickering, 1833), https://archive.org/details/astronogenphysics00whewuoft.

18. William Prout, Bridgewater Treatise no. 8, *Chemistry, Meteorology, and the Function of Digestion Considered with Reference to Natural Theology* (London: William Pickering, 1834), 440, https://archive.org/stream/chemistrymeteoro00pro#page/n19/mode/2up.

19. Friedrich Nietzsche, *The Will to Power* [1901], trans. Walter Kaufmann and R. J. Hollingdale (New York: Vintage Books, 1968).

20. Alfred Russel Wallace, *The World of Life: A Manifestation of Creative Power, Directive Mind and Ultimate Purpose* (London: Chapman and Hall, 1911).

21. Henderson, *Fitness*, 89.

22. Henderson, *Fitness*, 102.

23. Henderson, *Fitness*, 139–134.

24. George Wald, "The Origins of Life," *Proceedings of the National Academy of Sciences* 52, no. 2 (August 1, 1964): 595–611, https://doi.org/10.1073/pnas.52.2.595.

25. Harold J. Morowitz, *Cosmic Joy and Local Pain: Musings of a Mystic Scientist* (New York: Scribner, 1987), 152–153.

26. Wald, "The Origins of Life."

27. George Wald, "Light and Life," *Scientific American* 201, no. 4 (1959): 92–108.

28. Morowitz, *Cosmic Joy and Local Pain*.

29. Freeman Dyson, *Disturbing the Universe* (New York: Basic Books, 1979), 250.

30. John N. Maina, "Comparative Respiratory Morphology: Themes and Principles in the Design and Construction of the Gas Exchangers," *Anatomical Record* 261 (2000): 26.

31. "Blood Vessels," Atlas of Human Cardiac Anatomy, University of Minnesota, 2021, http://www.vhlab.umn.edu/atlas/physiology-tutorial/blood-vessels.shtml.

32. Narla Mohandas and Patrick G. Gallagher, "Red Cell Membrane: Past, Present, and Future," *Blood* 112, no. 10 (November 15, 2008): 3939–3948, https://doi.org/10.1182/blood-2008-07-161166.

33. Norman L. Jones, *The Ins and Outs of Breathing: How We Learnt About the Body's Most Vital Function* (Bloomington, IN: iUniverse, 2011), 35.

34. William Harvey, *De Motu Cordis* [1628], https://archive.org/details/onmotionheartan-00harvgoog/page/n4

35. John Hudson Tiner, *Exploring the History of Medicine: From the Ancient Physicians of Pharaoh to Genetic Engineering* (Green Forest, AR: Master Books, 2001), 36.

36. M. F. Perutz et al., "Structure of Hæmoglobin: A Three-Dimensional Fourier Synthesis at 5.5-Å. Resolution, Obtained by X-Ray Analysis," *Nature* 185, no. 4711 (February 13, 1960): 416–422, https://doi.org/10.1038/185416a0.

37. Mohandas and Gallagher, "Red Cell Membrane: Past, Present, and Future."

38. Nick Lane, *The Vital Question: Why Is Life the Way It Is?* (London: Profile Books, 2015), 75–77.

2. Prior Fitness: The Hydrological Cycle

1. Richard Bentley, *The Folly and Unreasonableness of Atheism* (London: J. H. for H. Mortlock, 1699), 265, https://archive.org/stream/follyandunreaso00bentgoog#page/n271/mode/2up/.

2. Philip Ball, H_2O: *A Biography of Water* (London: Weidenfeld and Nicolson, 1999), 26.

3. Even CO_2, which freezes at –78 C, exists on Earth only as a gas, as the partial pressure of CO_2 is so low that the rate of sublimation greatly exceeds the rate of condensation. See Anthony Watts, "Results: Lab Experiment Regarding CO_2 'Snow' in Antarctica at –113 F (–80 C) —Not Possible," Watts Up With That?, June 13, 2009, https://wattsupwiththat.com/2009/06/13/results-lab-experiment-regarding-co2-snow-in-antarctica-at-113f-80-5c-not-possible/.

4. Ball, H_2O, 23–24.

5. Lawrence J. Henderson, *The Fitness of the Environment: An Enquiry into the Biological Significance of the Properties of Matter* (New York: McMillan, 1913), 114.

6. Felix Franks, "Water the Unique Chemical," in *Water: A Comprehensive Treatise*, ed. Felix Franks, vol. 1 (New York: Plenum Press, 1972), 20.

7. Henderson, *Fitness*, 126.

8. Note that water expands when it freezes because of its unique hydrogen-bonded network. As the temperature cools towards freezing the water molecules are organized into more perfect, less mobile hexagonal-shaped hydrogen-bonded networks. Their more open nature is why ice is less dense than liquid water.

9. Only one other substance, the metal gallium, expands on freezing in the ambient temperature range. N. N. Greenwood and A. Earnshaw, *Chemistry of the Elements*, 2nd ed. (Oxford, UK: Butterworth-Heinemann, 1997), 223–224.

10. Marcia Bjornerud, *Reading the Rocks: The Autobiography of Earth* (New York: Basic Books, 2005), 73.

11. Glenn Elert, "Viscosity," Physics Hypertextbook, 2017, http://physics.info/viscosity/.

12. Shi Yaolin and Cao Jianling, "Lithosphere Effective Viscosity of Continental China," *Earth Science Frontiers* 15, no. 3 (May 2008): 82–95, https://doi.org/10.1016/S1872-5791(08)60064-0.

13. Bjornerud, *Reading the Rocks*, 73.

14. "Niagara Falls," Geology, New York State Museum, 2017, accessed March 4, 2022, http://www.nysm.nysed.gov/research-collections/geology/resources/niagara-falls.

15. Gerard V. Middleton and Peter R. Wilcock, *Mechanics in the Earth and Environmental Sciences* (Cambridge, UK: Cambridge University Press, 1994), 84.

16. Michael Pidwirny writes, "Because of basal sliding, some glaciers can move up to 50 meters in one day. However, average rates of movement are usually less than 1 meter per day." Michael Pidwirny, "Glacial Processes," *Fundamentals of Physical Geography*, 2nd ed. (2006), http://www.physicalgeography.net/fundamentals/10ae.html.

17. Nyle C. Brady and Raymond Weil, *Elements of the Nature and Properties of Soils*, 14th ed. (Harlow, Essex: Pearson Education Limited, 2016), 22.

18. Nyle C. Brady and Raymond R. Weil, *The Nature and Properties of Soils* (New Jersey: Prentice Hall, 1996), 242.

19. Brady and Weil, *The Nature and Properties of Soils*, 241.

20. Brady and Weil, *The Nature and Properties of Soils*, 270.

21. Ball, H_2O, 27.

22. As J. C. I. Dooge points out, by carefully calculating the amount of rainfall from several gauges in England and Wales and estimating the total runoff over the same areas, Dalton was able to show that rainfall was sufficient to supply rivers and streams with water. Dalton concluded, "Upon the whole then I think that we can finally conclude that the rain and dew of this country are equivalent to the quantity carried off by evaporation and the rivers" (8). Quoted in J. C. I .Dooge, "Concepts of the Hydrological Cycle, Ancient and Modern," International Symposium on H_2O: 'Origins and History of Hydrology,'" Dijon, May 9–11, 2001, http://hydrologie.org/ACT/OH2/actes/03_dooge.pdf. According to Dooge, Dalton's 1799 paper was reprinted three years later in Volume 5 of the Memoirs of the Society.

23. Yi-Fu Tuan, *The Hydrologic Cycle and the Wisdom of God* (Toronto: University of Toronto Press, 1968), 130.

24. Tuan, *The Hydrologic Cycle*, 4–6.

25. John Ray, *The Wisdom of God Manifested in the Works of the Creation* (London: W&J Innys, 1743), 89, https://archive.org/details/wisdomofgodmanif00ray.

26. Ray, *The Wisdom of God*, 90.

27. Tuan, *The Hydrological Cycle*, 4.

28. Tuan, *The Hydrological Cycle*, 4–6.

3. FITNESS FOR AEROBIC LIFE

1. Nick Lane, *Oxygen: The Molecule That Made the World* (Oxford: Oxford University Press, 2002), 131.

2. Kate Jackson, "As David Blaine Breaks the Record for Holding His Breath We Find More Superhuman Feats," *Mirror*, May 2, 2008, http://www.mirror.co.uk/news/uk-news/as-david-blaine-breaks-the-record-for-holding-305724.

3. Roman Boulatov, "Understanding the Reaction That Powers This World: Biomimetic Studies of Respiratory O_2 Reduction by Cytochrome Oxidase," *Pure and Applied Chemistry* 76 no. 2 (2004): 303–319, https://doi.org/10.1351/pac200476020303.

4. Norman R. Pace, "A Molecular View of Microbial Diversity and the Biosphere," *Science* 276, no. 5313 (May 2, 1997): 734–740, https://doi.org/10.1126/science.276.5313.734; William J. Broad, *The Universe Below: Discovering the Secrets of the Deep Sea* (New York: Simon & Schuster, 1997). See also Michael Denton, *Children of Light: The Astonishing Properties of Sunlight That Make Us Possible* (Seattle, WA: Discovery Institute Press, 2018), appendices "Doing without Oxygen" and "Respiration and Fermentation."

5. George Wald, "The Origin of Life," *Scientific American* 191, no. 2 (August 1954): 44–53. Wald wrote, "It is difficult to overestimate the degree to which the invention of cellular respiration [oxidation] released the forces of living organisms… No organism that relies wholly on fermentation has ever amounted to much."

6. Note another reason that fluorine is unsuitable is that the fluorine carbon bonds—as occur in the chemically inert and heat resistant polymer Teflon—are too strong to be broken or manipulated by biochemical systems.

7. David C. Catling et al., "Why O_2 Is Required by Complex Life on Habitable Planets and the Concept of Planetary 'Oxygenation Time,'" *Astrobiology* 5, no. 3 (June 7, 2005), https://doi.org/10.1089/ast.2005.5.415.

8. Sara Seager and William Bains, "The Search for Signs of Life on Exoplanets," *Science Advances* 1, no. 2 (March 6, 2015), https://doi.org/10.1126/sciadv.1500047; Lyn J. Rothschild, "The Evolution of Photosynthesis... Again?," *Philosophical Transactions of the Royal Society B: Biological Sciences* 363, no. 1504 (August 27, 2008): 2787–2801, https://doi.org/10.1098/rstb.2008.0056.

9. A. Léger et al., "Is the Presence of Oxygen on an Exoplanet a Reliable Biosignature?" *Astrobiology* 11, no. 4 (May 2011): 335–341, https://doi.org/10.1089/ast.2010.0516.

10. C. R. Nave, "The Interaction of Radiation with Matter," HyperPhysics, Module 3, Department of Physics and Astronomy, Georgia State University, accessed March 5, 2022, http://hyperphysics.phy-astr.gsu.edu/hbase/mod3.html.

11. George Wald, "Light and Life," *Scientific American* 201, no. 4 (1959): 92–108. Wald comments that radiations of wavelengths shorter than 300 millimicrons "are incompatible with the orderly existence of such large, highly organized molecules as proteins and nucleic acids. Both types of molecule consist of long chains of units bound together by primary valences [ordinary chemical bonds]. Both types of molecule, however, are held in their delicate and specific configurations upon which their functions in the cell depend by the relatively weak forces of hydrogen-bonding and van der Waals attraction. These forces, though individually weak, are cumulative. They hold a molecule together in a specific arrangement, like zippers. Radiation of wavelengths shorter than 300 millimicrons unzips them, opening up long sections of attachment, and permitting the orderly arrangement to become random and chaotic. Hence such radiations denature proteins and depolymerize nucleic acids with disastrous consequences for the cell. For this reason about 300 millimicrons represents the lower limit of radiation capable of promoting photoreactions, yet compatible with life."

12. Denton, *Children of Light*, 22.

13. Geoffrey K. Vallis, *Climate and the Oceans* (Princeton, NJ: Princeton University Press, 2012), 218.

14. John F. B. Mitchell, "The 'Greenhouse' Effect and Climate Change," *Reviews of Geophysics*, 27, no. 1 (February 1989), https://doi.org/10.1029/RG027i001p00115.

15. Mitchell, "The 'Greenhouse' Effect and Climate Change."

16. I. M. Campbell, *Energy and the Atmosphere* (London: Wiley, 1977), 1–2.

17. Campbell, *Energy and the Atmosphere*, 2.

18. Denton, *Children of Light*, 20.

19. Carlos A. Bertulani, *Nuclei in the Cosmos* (Hackensack, NJ: World Scientific, 2013), chap. 1.

20. "Colors, Temperatures, and Spectral Types of Stars," John A. Dutton e-Education Institute, Penn State, 2017, accessed January 11, 2018, https://www.e-education.psu.edu/astro801/content/l4_p2.html.

21. Vallis, *Climate and the Oceans*, 218. Vallis writes that without the greenhouse effect, "Earth's surface would have a temperature of about 255K (–18C), about 33 degrees lower than it actually is."

22. David R. Williams, "Moon Fact Sheet," Lunar and Planetary Science, NASA, December 20, 2021, https://nssdc.gsfc.nasa.gov/planetary/factsheet/moonfact.html. The diurnal temperature range on the moon (at the equator) is from 95°K (–178°C) to 390 K (116 C).

23. Mitchell, "The 'Greenhouse' Effect and Climate Change."

24. Mitchell, "The 'Greenhouse' Effect and Climate Change."

25. Stephen G. Warren, Richard E. Brandt, and Thomas C. Grenfell, "Visible and Near-Ultraviolet Absorption Spectrum of Ice from Transmission of Solar Radiation into Snow," *Applied Optics* 45, no. 21 (July 20, 2006): 5320, https://doi.org/10.1364/. AO.45.005320. The authors comment, "The general features of the spectrum are well known… Ice exhibits strong absorption in the ultraviolet (UV) at wavelength170 nm. With increasing wavelength, the absorption becomes extremely weak in the visible, with a minimum near 400 nm." See also Stephen G. Warren and Richard E. Brandt, "Optical Constants of Ice from the Ultraviolet to the Microwave: A Revised Compilation," *Journal of Geophysical Research* 113, no. D14 (July 31, 2008), https://doi.org/10.1029/2007JD009744.

26. Denton, *Children of Light*, 52. For more detail and references on this matter, see Chapter 3 of *Children of Light*.

27. Carl Sagan, *Cosmos* (New York: Ballantine Books, 1980), 80.

28. Dana L. Royer, "CO_2-Forced Climate Thresholds during the Phanerozoic," *Geochimica et Cosmochimica Acta* 70, no. 23 (December 2006): 5665–5675, https://doi.org/10.1016/j.gca.2005.11.031.

29. Joseph G. Allen et al., "Associations of Cognitive Function Scores with Carbon Dioxide, Ventilation, and Volatile Organic Compound Exposures in Office Workers: A Controlled Exposure Study of Green and Conventional Office Environments," *Environmental Health Perspectives* 124, no. 6 (October 26, 2015), https://doi.org/10.1289/ehp.1510037. See also Edward W. Schwieterman et al., "A Limited Habitable Zone for Complex Life," *The Astrophysical Journal* 878, no. 19 (June 10, 2019), https://iopscience.iop.org/article/10.3847/1538-4357/ab1d52.

30. Marcia Bjornerud, *Reading the Rocks* (Cambridge, MA: Westview Press, 2005). As she points out in Chapter 1, the temperature of the surface of Venus is 806°F (430°C). The melting point of lead is 630°F (332°C).

31. "Ozone," Climate Changers, accessed March 18, 2022, http://www.change-climate.com/GreenHouse_Gases/Ozone.htm.

32. "Electromagnetic Radiation," in *Encyclopedia Britannica*, 15th ed. (1994), vol. 18, 203.

33. William J. Broad, *The Universe Below: Discovering the Secrets of the Deep Sea* (New York: Simon & Schuster, 1997).

4. Prior Fitness: The Atmosphere

1. Nick Lane, *The Vital Question: Why Is Life the Way It Is?* (London: Profile Books, 2015), 64.

2. Knut Schmidt-Nielsen, *Animal Physiology: Adaptation and Environment*, 5th ed. (Cambridge, UK: Cambridge University Press, 1997), chap. 1.

3. Robert C. Lasiewski, "Body Temperatures, Heart and Breathing Rate, and Evaporative Water Loss in Hummingbirds," *Physiological Zoology* 37, no. 2 (April 1964): 212–223, https://doi.org/10.1086/physzool.37.2.30152332.

4. R. K. Suarez, "Oxygen and the Upper Limits to Animal Design and Performance," *The Journal of Experimental Biology* 201, no. 8 (April 1998): 1065–1072.

5. Suarez, "Oxygen and the Upper Limits." See also "P/O Ratio," A Dictionary of Biology, Encyclopedia.com, https://www.encyclopedia.com/science/dictionaries-thesauruses-pictures-and-press-releases/po-ratio; and P. C. Hinkle and M. L. Yu, "The Phosphorus/Oxygen Ratio of Mitochondrial Oxidative Phosphorylation," *The Journal of Biological Chemistry* 254, no. 7 (April 10, 1979): 2450–2455.

6. John N. Maina, "Structure, Function and Evolution of the Gas Exchangers: Comparative Perspectives," *Journal of Anatomy* 201, no. 4 (October 2002): 284.

7. Goran E. Nilsson, ed., *Respiratory Physiology of Vertebrates: Life with and without Oxygen* (New York: Cambridge University Press, 2010), 5.

8. J. B. Graham and K. A. Dickson, "Tuna Comparative Physiology," *Journal of Experimental Biology* 207, no. 23 (November 1, 2004): 4015–4024, https://doi.org/10.1242/jeb.01267. The authors comment that tunas approach "the functional limits imposed by physical and biological principles." Additional increases in the metabolic rate and the consumption of oxygen are probably limited by "diminishing returns." For instance, increasing gill area, although that might improve oxygen extraction marginally, would tend to increase both heat loss and the metabolic cost of regulating ion fluxes across the gill membranes. Swimming faster to increase ventilation of the gills would require additional muscular power to propel the fish through the water faster, additional muscular power that in turn would exact higher aerobic costs.

9. Using conversion rates given at https://www.aqua-calc.com/calculate/volume-to-weight, metabolic rate in resting tuna given as 12.25 mg /Kg/min in A. P. Farrell, "From Hagfish to Tuna: A Perspective on Cardiac Function in Fish," *Physiological Zoology* 64 (1991):1137–1164; resting human given as 7.15 mg /kg /min in John N. Maina, "Comparative Respiratory Morphology: Themes and Principles in the Design and Construction of the Gas Exchangers," *The Anatomical Record* 261 (2000): 25–44, 26. Max tuna given as 42 mg /kg/ min in Graham and Dickson, "Tuna Comparative Physiology," and in R. W. Brill and P. G. Bushnell, "The Cardiovascular System of Tunas," *Tuna: Physiology, Ecology and Evolution*, vol. 19, eds. B. A. Block and E. D. Stevens (San Diego: Academic Press, 2001), 79–120. Max human given as 171 mg /kg/min in Maina, "Comparative Respiratory Morphology." Above estimates inferred also from data in Suarez, "Oxygen and the Upper Limits to Animal Design." Note a recent paper reports a fish with the highest metabolic rate yet recorded. See Gudrun De Boeck et al., "Mammalian Metabolic Rates in the Hottest Fish on Earth," *Scientific Reports* 6, article number 26990 (July 2016), https://doi.org/10.1038/srep26990. The authors comment, "It is impressive that this fish can sustain an MO_2 (max) equal to about one-third of the MO_2 (max) (507 μmol g–1 h–1) of the pygmy mouse."

10. Gudrun De Boeck et al., "Metabolic Rates." The exception is a fish, the opah, which was shown to have achieved whole body endothermy, being able to maintain its whole body temperature 5°C above the surrounding water. But this differential pales against the differential maintained in mammals and birds. Whales, seals, and penguins in the polar waters maintain body temperatures nearly 40°C above their surrounding sea, and a polar bear overwintering on the frozen Arctic ocean has a body temperature 80°C above its surroundings.

Warm-blooded endothermy comparable with that of birds and mammals is simply beyond the reach of organisms that breathe through gills. The metabolic cost in water is too great.

11. Goren E. Nilsson, "Brain and Body Oxygen Requirements of Gnathonemus Petersii, a Fish with an Exceptionally Large Brain," *The Journal of Experimental Biology* 199, no. 3 (1996): 603–607. African electric fish are also reported to engage in play, a sign of intelligence common among warm-blooded animals. Gene F. Helfman et al., *The Diversity of Fishes: Biology, Evolution, and Ecology*, 2nd ed. (Hoboken, NJ: Blackwell, 2009), 6, 263.

12. Claude Yoder, "Solubility of Gases in Water at 293K," Wired Chemist, 2022, https://www.wiredchemist.com/chemistry/data/solubilities-gases; "Solubility of Gases in Water vs. Temperature," Engineering Toolbox, 2008, https://www.engineeringtoolbox.com/gases-solubility-water-d_1148.html.

13. David C. Catling et al., "Why O_2 Is Required by Complex Life on Habitable Planets and the Concept of Planetary 'Oxygenation Time,'" *Astrobiology* 5, no. 3 (June 2005): 415–438, https://doi.org/10.1089/ast.2005.5.415.

14. At the summit of Mount Everest, the partial pressure of oxygen is only 53 mm Hg compared with 160 mm Hg at sea level. See John B. West et al., "Barometric Pressures at Extreme Altitudes on Mt. Everest: Physiological Significance," *Journal of Applied Physiology* 54, no. 5 (May 1983): 1188–1194, https://doi.org/10.1152/jappl.1983.54.5.1188; John B. West, "Barometric Pressures on Mt. Everest: New Data and Physiological Significance," *Journal of Applied Physiology* 86, no. 3 (March 1999): 1062–1066, https://doi.org/10.1152/jappl.1999.86.3.1062.

15. P. D. Wagner, "The Biology of Oxygen," *European Respiratory Journal* 31, no. 4 (April 1, 2008): 887–890, https://doi.org/10.1183/09031936.00155407.

16. John B. West, "Highest Permanent Human Habitation," *High Altitude Medicine & Biology* 3, no. 4 (June 6, 2004): 401–407, https://doi.org/10.1089/15270290260512882. PMID 12631426.

17. Phillip C. Withers et al., *Ecological and Environmental Physiology of Mammals*: Ecological and Environmental Physiology Series, vol. 5 (Oxford, UK: Oxford University Press, 2016), 330.

18. John B. West, "Highest Permanent Human Habitation," 401.

19. Tim Lenton, *Revolutions That Made the Earth* (Oxford, UK: Oxford University Press, 2011), chap. 15.

20. Emilia Huerta-Sánchez et al., "Altitude Adaptation in Tibetans Caused by Introgression of Denisovan-like DNA," *Nature* 512, no. 7513 (July 2, 2014): 194–197, https://doi.org/10.1038/nature13408; John B. West, "Human Responses to Extreme Altitudes," *Integrative and Comparative Biology* 46, no. 1 (January 6, 2006): 25–34, https://doi.org/10.1093/icb/icj005; Tianya Wu and Bengt Kayser, "High Altitude Adaptation in Tibetans," *High Altitude Medicine and Biology* 7, no. 3 (2006): 193–208, https://doi.org/10.1089/ham.2006.7.193; J. L. Rupert and P. W. Hochachka, "Genetic Approaches to Understanding Human Adaptation to Altitude in the Andes," *The Journal of Experimental Biology* 204, no. 18 (September 2001): 3151–3160.

21. Sara Seager and William Bains, "The Search for Signs of Life on Exoplanets," *Science Advances* 1, no. 2 (March 6, 2015), https//doi.org/10.1126/sciadv.1500047.

22. Roman Boulatov, "Understanding the Reaction That Powers This World: Biomimetic Studies of Respiratory O_2 Reduction by Cytochrome Oxidase," *Pure and Applied Chemistry* 76, no. 2 (2004): 303–319, https://doi.org/10.1351/pac200476020303.

23. *Mysterious Universe*, season 1, episode 8, "The Burning Question," Arthur C. Clarke, aired March 8, 1994.

24. Nick Lane, *Oxygen: The Molecule That Made the World* (Oxford, UK: Oxford University Press, 2002), 119.

25. Lane, *Oxygen*, 121. Bracketed English system quantities in original.

26. Boulatov, "Understanding the Reaction That Powers This World."

27. Alfred Russel Wallace, *The World of Life* (London: Chapman and Hall, 1914), chap. 18, https://archive.org/details/worldoflifemanif00walliala/page/362. Wallace writes, "In Chamber's Encyclopedia we find the following statement: 'At ordinary temperatures all the varieties of carbon are extremely unalterable; so much so that it is customary to burn the ends of piles of wood which are to be driven into the ground, so that the coating of non decaying carbon may preserve the inner wood.'"

28. William Prout, Bridgewater Treatise No. 8, *Chemistry, Meteorology, and the Function of Digestion Considered with Reference to Natural Theology* (London: William Pickering, 1834), 103, https://archive.org/stream/chemistrymeteoro00pro#page/n19/mode/2up.

29. Norman V. Sidgwick, *The Chemical Elements and their Compounds*, vol. 1 (Oxford, UK: Oxford University Press, 1950), 490. Sidgwick comments: "In the first place the typical 4-covalent state of the carbon atom is one in which all the formal elements of stability are combined. It has an octet, a fully shared octet, an inert gas number, and in addition unlike all the other elements of the group, an octet which cannot increase beyond 8, since 4 is the maximum covalency possible for carbon. Hence the saturated carbon atom cannot co-ordinate either as donor or as acceptor, and since by far the commonest method of reaction is through co-ordination, carbon is necessarily very slow to react and even in a thermodynamically unstable molecule may actually persist for a long time unchanged."

30. Lane, *Oxygen*, 10, 119.

31. M. J. Green and A. O. Hill, "The Chemistry of Dioxygen," *Methods in Enzymology* 105 (1984): 1–21; See also Ali Naqui, Britton Chance, and Enrique Cadenas, "Reactive Oxygen Intermediates in Biochemistry," *Annual Review of Biochemistry* 55 (1986): 137–166. As the authors note, "Oxygen is… the only element in the most appropriate physical state, with a satisfactory solubility in water and with desirable combinations of kinetic and thermodynamic properties." It isn't too reactive. At the same time, it isn't so inert that cellular processes cannot coax it into the reactions required of it as a prime energy source of life. Several recent scholarly reviews explain in detail why the oxygen atom's unique electronic structure confers upon oxygen its reluctance to react and how this reluctance is overcome by the cell. See the Green and Hill article cited immediately above, as well as Boutalov, "Understanding the Reaction that Powers this World"; Irwin Fridovich, "Oxygen: How Do We Stand It?," *Medical Principles and Practice* 22, no. 2 (2013): 131-137. "Oxygen: How Do We Stand It?"; and K. P. Jensen and U. Ryde, "How O_2 Binds to Heme: Reasons for Rapid Binding and Spin Inversion," *Journal of Biological Chemistry* 279, no. 15 (April 9, 2004): 14561–14569, https://doi.org/10.1074/jbc.M314007200. The authors comment, "All electrons have a spin, which is an intrinsic quantum chemical property that can take only two possible values, normally called alpha and beta (or spin up and down). Almost all normal organic molecules contain an even number of electrons and also an equal number of alpha and beta electrons. They are then said to have paired spin or to be singlets. Molecular oxygen (O_2) is a famous exception to this rule. In its ground state, it has two more electrons of one spin state than the other. Thus, it is said to have two unpaired electrons or to be a triplet. The singlet state of O_2, with all electron spins paired, is… [approximately] 90 kJ/mol higher in energy than the triplet ground state. A chemical reaction can normally not change the spin state of an

electron. Therefore, reactions between singlet and triplet states are formally spin-forbidden, which means that they are slow. This is the reason why organic matter may exist in an atmosphere containing much O_2. There is a strong thermodynamic drive of O_2 to oxidize organic matter to H_2O and CO_2 but because these products (as well as the organic molecules) are singlets (whereas O_2 is a triplet), this reaction is spin-forbidden and therefore very slow at ambient temperatures. On the other hand, this is a problem when living organisms want to employ O_2 in their metabolism; the reactions are still spin-forbidden and slow. Nature has handled this problem by using transition metals to carry, activate, and reduce O_2. There are many reasons for this choice. First, most transition metals also contain unpaired electrons, allowing reactions with triplet O_2. Second, transition metals are relatively heavy atoms, which increases spin-orbit coupling (SOC), and thereby provide a quantum mechanical mechanism to change the spin state of an electron, called spin inversion. However, the SOC of the first-row transition metals is too small alone to allow for spin transitions. Third, transition metals often have several excited states with unpaired electrons close in energy to the ground state. This can also be used to enhance the probability of spin inversion."

32. Boulatov, "Understanding the Reaction That Powers This World."

33. Douglas Drysdale, *An Introduction to Fire Dynamics*, 3rd ed. (Chichester, West Sussex: Wiley, 2011), 378.

34. Tim Lenton and Andrew Watson, *Revolutions That Made the Earth* (New York: Oxford University Press, 2011), chap. 15, 298.

35. Rodrigo Luger and R. Barnes, "Extreme Water Loss and Abiotic O_2 Buildup on Planets Throughout the Habitable Zones of M Dwarfs," *Astrobiology* 15, no. 2 (February 2015): 119–143, https://doi.org/10.1089/ast.2014.1231.

36. A. E. Needham, *The Uniqueness of Biological Materials* (London: Pergamon Press, 1965), 149.

37. Needham, *The Uniqueness of Biological Materials*, 149–150.

38. Sami Mikhail and Dimitri A. Sverjensky, "Nitrogen Speciation in Upper Mantle Fluids and the Origin of Earth's Nitrogen-Rich Atmosphere," *Nature Geoscience* 7 (October 19, 2014): 816.

39. Stuart Ross Taylor and Scott M. McLennan, *Planetary Crusts: Their Composition, Origin and Evolution*, Cambridge Planetary Science (Cambridge, UK: Cambridge University Press, 2009), chaps. 1 and 14; Shannon Hall, "Earth's Tectonic Activity May Be Crucial for Life— and Rare in Our Galaxy," *Scientific American* (July 20, 2017), https://www.scientificam-erican.com/article/earths-tectonic-activity-may-be-crucial-for-life-and-rare-in-our-galaxy/; David Waltham, *Lucky Planet: Why Earth Is Exceptional—And What That Means for Life in the Universe* (New York: Basic Books, 2014); Peter D. Ward, *Life as We Do Not Know It: The NASA Search for (and Synthesis of) Alien Life* (London: Penguin, 2007).

40. Edward T. McHale, "Habitable Atmospheres which Do Not Support Combustion," US Defense Technical Information Center, paper 30, (1972): 331–335, https://ia600301. us.archive.org/18/items/nasa_techdoc_19720014620/19720014620.pdf. Near identical text also available at E. T. McHale, "Life Support without Combustion Hazards," *Fire Technology* 10, no. 1 (1974): 15–24, https://doi.org/https://doi.org/10.1007/BF02590509.

41. McHale, "Habitable Atmospheres which Do Not Support Combustion."

42. McHale, "Habitable Atmospheres which Do Not Support Combustion."

43. Drysdale, *An Introduction to Fire Dynamics*, 378. See also Clayton Huggett, "Habitable Atmospheres which Do Not Support Combustion," *Flame and Combustion* 20, no. 1 (1972): 140–142; Vytenis Babrauskas and Stephen J. Grayson, eds., *Heat Release in Fires* (London:

Interscience Communications, 2009). The authors comment: "The effects of pressure on the burning of combustibles has become of great interest to the U.S. Navy as a means of extinguishing fires... In a landmark paper entitled 'Habitable Atmospheres which Do Not Support Combustion,' Huggett explained that for survival humans depend on there being a minimum partial pressure of oxygen, and a minimum concentration. By contrast, the combustion process requires a minimum flame temperature to avoid extinction. This minimum flame temperature can be related to a minimum heat capacity per mole of O_2, this being about 170 to 210 J/C per mole of O_2. Thus, if the total pressure of the atmosphere is increased by the forced injection of an inert gas into a sealed atmosphere, it may be possible to extinguish a fire without injuring persons. In small-scale pool fire tests, extinction was typically achieved when the nitrogen diluent raised the total pressure to about 1.6 atmospheres. Subsequent, engineering details have been pursued in an ambitious program of large scale tests" (316).

44. John Hunt, *The Ascent of Everest* (Seattle, WA: Mountaineers, 1998), 206. See also "Everest Anniversary: How Climbing the World's Highest Mountain Has Changed," *Telegraph*, May 29, 2013, http://www.telegraph.co.uk/news/worldnews/asia/mounteverest/10085905/Everest-anniversary-how-climbing-the-worlds-highest-mountain-has-changed.html.

5. BREATHING

1. Ewald R. Weibel, "What Makes a Good Lung?," *Swiss Medical Weekly* 139, no. 27–28 (July 11, 2009): 375–386, http://docs.wixstatic.com/ugd/0ead26_a07bc9dd49034303ab26ddd8d0ecfcfe.pdf.

2. Knut Schmidt-Nielsen, *Animal Physiology: Adaptation and Environment*, 5th ed. (New York: Cambridge University Press, 1997), chap. 1; John N. Maina, "Structure, Function and Evolution of the Gas Exchangers: Comparative Perspectives," *Journal of Anatomy* 201, no. 4 (October 2002): 281–304.

3. John Maina, "Structure, Function and Evolution of the Gas Exchangers: Comparative Perspectives," *Journal of Anatomy* 201, no. 4 (October 2002).

4. Maina, "Structure, Function and Evolution of the Gas Exchangers."

5. James P. Butler and Akira Tsuda, "Transport of Gases between the Environment and Alveoli—Theoretical Foundations," *Comprehensive Physiology* 1, no. 3 (July 2011): 1301–1316. https://doi.org/10.1002/cphy.c090016. Also see James S. Ultman, *Air Pollution, the Automobile, and Public Health* (Washington DC: National Academies Press, 1988), https://www.ncbi.nlm.nih.gov/books/NBK218139/; and Maina, "Structure, Function and Evolution of the Gas Exchangers."

6. Butler and Tsuda, "Transport of Gases." Note that diffusion is very effective over short distances in the respiratory zone and cost free, but it is ineffective over long distances in the conducting zone. See Schmidt-Nielsen, *Animal Physiology*, 585.

7. D. S. Minors, "Abnormal Pressure," in *Variations in Human Physiology*, ed. R. M. Case (Manchester, UK: Manchester University Press, 1985), 105.

8. S. Zakynthinos and C. Roussos, "Oxygen Cost of Breathing," *Tissue Oxygen Utilization, Update in Intensive Care and Emergency Medicine* Series vol. 12, (Berlin, Heidelberg: Springer Berlin Heidelberg, 1991), 171–184, https://doi.org/10.1007/978-3-642-84169-9_14.

9. Craig A. Harms et al., "Effects of Respiratory Muscle Work on Exercise Performance," *Journal of Applied Physiology* 89, no. 1 (July 2000): 131–138, https://doi.org/10.1152/jappl.2000.89.1.131. Posted at https://www.physiology.org/doi/full/10.1152/jappl.2000.89.1.131. As the authors comment, "Respiratory musculature during strenuous

exercise in humans can command ~10% of the total oxygen consumption… in moderately fit subjects and up to 15% in highly fit subjects." See also E. A. Aaron et al., "Oxygen Cost of Exercise Hyperpnea: Measurement," *Journal of Applied Physiology* 72, no. 5 (May 1992): 1810–1817, https://doi.org/10.1152/jappl.1992.72.5.1810.

10. Giovanni Vladilo et al., "The Habitable Zone of Earth-like Planets with Different Levels of Atmospheric Oxygen," *The Astrophysical Journal* 767, no. 1 (April 10, 2013): 65, https://doi.org/10.1088/0004-637X/767/1/65.

11. R. Luger and R. Barnes, "Extreme Water Loss and Abiotic O_2 Buildup on Planets Throughout the Habitable Zones of M Dwarfs," *Astrobiology* 15, no. 2 (February 2015): 119–143, https://doi.org/10.1089/ast.2014.1231. Keiko Hamano, Yutaka Abe, and Hidenori Genda extended this idea to exoplanetary systems, arguing for the existence of two fundamentally different types of terrestrial planets: type I planets, which undergo short-lived runaway greenhouses during their formation, and type II planets, which form interior to a critical distance and can remain in runaway greenhouses for as long as 100 Myr. The former type of planet, like Earth, retains most of its water inventory and may evolve to become habitable. The latter, similarly to Venus, undergoes complete surface desiccation during the runaway. Keiko Hamano, Yutaka Abe, and Hidenori Genda, "Emergence of Two Types of Terrestrial Planet on Solidification of Magma Ocean," *Nature* 497, no. 7451 (May 2013): 607–610, https://doi.org/10.1038/nature12163. See also Colin Goldblatt and Andrew J. Watson, "The Runaway Greenhouse: Implications for Future Climate Change, Geoengineering and Planetary Atmospheres," *Philosophical Transactions of the Royal Society A: Mathematical, Physical and Engineering Sciences* 370, no. 1974 (September 13, 2012): 4197–4216, https://doi.org/10.1098/rsta.2012.0004.

12. Carl R. Nave, "Viscosity of Liquids and Gases," *Hyperphysics*, Georgia State University, accessed March 10, 2022, http://hyperphysics.phy-astr.gsu.edu/hbase/Tables/viscosity.html.

13. Maina, "Structure, Function and Evolution of the Gas Exchangers."

14. "Effect of Moisture Content on the Viscosity of Honey at Different Temperatures," *Journal of Food Processing* 72, no. 4 (February 2006), 372–377.

15. Maina, "Structure, Function and Evolution of the Gas Exchangers."

16. J. A. Kylstra, C. V. Paganelli, and E. H. Lanphier, "Pulmonary Gas Exchange in Dogs Ventilated with Hyperbarically Oxygenated Liquid," *Journal of Applied Physiology* 21 (1996): 177–184; Maria Laura Costantino et al., "Clinical Design Functions: Round Table Discussions on the Bioengineering of Liquid Ventilators," *ASAIO Journal* 55, no. 3 (May 2009), 206–208, https://doi.org/10.1097/MAT.0b013e318199c167.

17. Steven Vogel, *Comparative Biomechanics: Life's Physical World*, 2nd ed. (Princeton, NJ: Princeton University Press, 2013), chap. 9; Schmidt-Nielsen, *Animal Physiology*, chap. 3 section "Viscosity."

18. Vogel, *Comparative Biomechanics*, 167.

19. Nave, "Viscosity of Liquids and Gases."

20. "Mercury," *Royal Society of Chemistry*, https://www.rsc.org/periodic-table/element/80/mercury; "Viscosity of Liquids and Gases, *Royal Society of Chemistry*, http://hyperphysics.phy-astr.gsu.edu/hbase/Tables/viscosity.html.

21. "Is There a Relationship between Viscosity and Density?," *Reference*, April 2, 2020, https://www.reference.com/science/relationship-between-viscosity-density-854fb3dae-786fff3.

22. "The Process of Breathing," *Lumen Learning*, accessed March 11, 2022, https://courses.lumenlearning.com/cuny-kbcc-ap2/chapter/the-process-of-breathing-no-content/.

23. Carl R. Nave, "Breathing Pressure Model," *Hyperphysics*, Georgia State University, accessed March 11, 2022, http://hyperphysics.phy-astr.gsu.edu/hbase/Kinetic/henry.html.

24. Weibel, "What Makes a Good Lung?"

25. Maina, "Structure, Function and Evolution of the Gas Exchangers." Also see "Diffusion," *Britannica*, accessed March 11, 2022, https://www.britannica.com/science/gas-state-of-matter/Diffusion.

26. Weibel, "What Makes a Good Lung?"

27. Matthias Ochs et al., "The Number of Alveoli in the Human Lung," *American Journal of Respiratory and Critical Care Medicine* 169, no. 1 (January 2004): 120–124, https://doi.org/10.1164/rccm.200308-1107OC.

28. "Color and Film Thickness," *Fandom*, accessed March 11, 2022, https://soapbubble.fandom.com/wiki/Color_and_Film_Thickness.

29. Ewald R. Weibel, "Lung Morphometry: The Link between Structure and Function," *Cell and Tissue Research* 367, no. 3 (March 2017): 413–426, https://doi.org/10.1007/s00441-016-2541-4.

30. Carl R. Nave, *Hyperphysics*, Georgia State University, accessed March 11, 2022, http://hyperphysics.phy-astr.gsu.edu/hbase/Kinetic/henry.html.

31. Wanda M. Haschek, Colin G. Rousseaux, and Matthew A. Wallig, "Respiratory System," *Fundamentals of Toxicologic Pathology* (London: Academic Press, 2010), 93–133, https://doi.org/10.1016/B978-0-12-370469-6.00006-4.

32. Schmidt-Nielsen, *Animal Physiology*, chap. 1, 29.

33. Weibel, "Lung Morphometry."

34. Weibel, "What Makes a Good Lung?"

35. Weibel, "What Makes a Good Lung?," 379.

36. Weibel, "What Makes a Good Lung?," 375.

37. Maina, "Comparative Respiratory Morphology."

38. Weibel, "What Makes a Good Lung?," 375. Another piece of evidence further highlighting the adaptive excellence of the barrier for gaseous exchange is that gaseous exchange is perfusion limited and not diffusion limited. This means that the 70 ml of blood pumped by the right ventricle through the pulmonary capillaries every second leaves the lungs fully oxygenated and, indeed, the pO_2 of the blood in the alveoli capillaries is fully oxygenated in only a fraction of the transit time even in strenuous exercise when the blood is fully oxygenated after only 0.25 seconds. See Joseph Feher, "Gas Exchange in the Lungs," *Quantitative Human Physiology* (Amsterdam: Elsevier, 2017), 642–652, https://doi.org/10.1016/B978-0-12-800883-6.00062-8.

39. See Carl R. Nave, *Hyperphysics*, Georgia State University, accessed March 11, 2022, http://hyperphysics.phy-astr.gsu.edu/hbase/Kinetic/henry.html In addition to Fick's law, two other laws relevant to this discussion include Henry's Law, which states that the amount of dissolved gas (in a liquid in contact with a gas) is proportional to its partial pressure in the gas phase; and Graham's Law, which states that the relative rate of diffusion of a gas in a liquid is proportional to its solubility and inversely proportional to the square root of its molecular weight.

40. "Oxygen Cascade," Life in the Fastlane, last modified August 23, 2021, https://partone.litfl.com/oxygen_cascade.html#id.

41. "Oxygen Cascade."

42. Carl R. Nave, "Saturated Vapor Pressure, Density for Water," Hyperphysics, Georgia State University, accessed March 11, 2022, http://hyperphysics.phy-astr.gsu.edu/hbase/Kinetic/watvap.html.

43. Calculated from formula given by Sandeep Sharma and Muhammad F. Hashmi, "Partial Pressure of Oxygen," National Center for Biotechnology Information, last modified September 30, 2021, https://www.ncbi.nlm.nih.gov/books/NBK493219/.

44. "Oxygen Cascade."

45. J. B. West, "Human Responses to Extreme Altitudes," *Integrative and Comparative Biology* 46, no. 1 (January 6, 2006): 25–34, https://doi.org/10.1093/icb/icj005.

46. Nave, "Saturated Vapor Pressure, Density for Water."

47. Geoffrey K. Vallis, *Climate and the Oceans*, Princeton Primers in Climate (Princeton, NJ: Princeton University Press, 2012), 218. Vallis writes, that without the greenhouse effect, "Earth's surface would have a temperature of about 255 K (–18°C), about 33 degrees lower than it actually is."

48. Lawrence J. Henderson, *The Fitness of the Environment: An Enquiry into the Biological Significance of the Properties of Matter* (New York: McMillan, 1913), 139–140.

49. The respiratory quotient of a typical western diet is generally cited as 0.8. See "Oxygen Cascade."

50. "Oxygen Cascade."

51. Chris Nickson, "Oxygen-Haemoglobin Dissociation Curve," Life in the Fastlane, November 3, 2020, https://litfl.com/oxygen-haemoglobin-dissociation-curve/.

52. Zachary Messina and Herbert Patrick, "Partial Pressure of Carbon Dioxide," National Center for Biotechnology Information, last modified September 28, 2021, https://www.ncbi.nlm.nih.gov/books/NBK551648/.

53. "Oxygen Cascade."

54. "Oxygen Cascade."

55. Joseph Feher, "Gas Exchange in the Lungs," *Quantitative Human Physiology* (Amsterdam: Elsevier, 2017), 642–652, https://doi.org/10.1016/B978-0-12-800883-6.00062-8.

56. Maina, "Structure, Function and Evolution of the Gas Exchangers."

57. Nave, http://hyperphysics.phy-astr.gsu.edu/hbase/Kinetic/henry.html.

58. P. H. Irving, *The Physics of the Human Body: A Physical View of Physiology Biological and Medical Physics, Biomedical Engineering* (New York: Springer, 2007).

59. Nave, http://hyperphysics.phy-astr.gsu.edu/hbase/Kinetic/henry.html.

60. "Solubilities of Gases in Water at 293 K," Wired Chemist, https://www.wiredchemist.com/chemistry/data/solubilities-gases; "Solubility of Gases in Water vs. Temperature," The Engineering Toolbox, https://www.engineeringtoolbox.com/gases-solubility-water-d_1148.html.

61. "Solubilities of Gases in Water at 293 K," Wired Chemist; "Solubility of Gases in Water vs. Temperature," The Engineering Toolbox.

62. Nave, http://hyperphysics.phy-astr.gsu.edu/hbase/Kinetic/henry.html.

63. Roland N. Pittman, *Regulation of Tissue Oxygenation* (San Rafael, CA: Morgan & Claypool Life Sciences, 2011), chap. 4, https://www.ncbi.nlm.nih.gov/books/NBK54103/.

64. Alex Yartsev, "Oxygen Carrying Capacity of Whole Blood," Deranged Physiology, June 8, 2015, https://derangedphysiology.com/main/cicm-primary-exam/required-reading/respiratory-system/Chapter%201111/oxygen-carrying-capacity-whole-blood. The solubility constant at 37°C is around 0.03ml/L/mmHg, or 0.003 ml/dL/mmHg. Thus, in every liter

of maximally oxygen-saturated blood (i.e., at an alveolar O_2 of around 100 mm Hg) there is only 3ml/L of dissolved oxygen, whereas there might be about 150-200ml/L of oxygens stored in hemoglobin.

65. David C. Catling et al., "Why O_2 Is Required by Complex Life on Habitable Planets and the Concept of Planetary 'Oxygenation Time,'" *Astrobiology* 5, no. 3 (June 2005): 415–438, https://doi.org/10.1089/ast.2005.5.415.

66. Irwin Fridovich, "Oxygen Toxicity: A Radical Explanation," *Journal of Experimental Biology* 201 (1998): 1203–1209. Maina, "Structure, Function and Evolution of the Gas Exchangers." Maina comments, "The assault by the RORs [reactive oxygen radicals] on the DNA, proteins and other macromolecules is profound. It is estimated, for example, that about 2–3% of oxygen taken up by aerobic cells results in production of O_2^- [superoxide anion] radical and H_2O_2 [hydrogen peroxide]; about 10^{12} oxygen molecules are handled by a rat cell daily, generating about 2 x 10^{10} (i.e. 2%) and H_2O_2; about 9 x 10^4 attacks on the DNA per day per cell occur in a rat; and RORs are responsible for 10,000 or so DNA base modifications per cell per day."

67. T. P. A. Devasagayam et al., "Free Radicals and Antioxidants in Human Health: Current Status and Future Prospects," *Journal of the Association of Physicians of India* 52 (2004):794–804. Chemical damage results from their uncontrolled oxidizing reactions with the organic molecules in the cell. And see A. Chawla and A. K. Lavania, "Oxygen Toxicity," *Medical Journal, Armed Forces India* 57, no. 2 (April 2001): 131–133, https://doi.org/10.1016/S0377-1237(01)80133-7. The authors comment thus: "Free radicals are defined as ions, atoms or molecules having an unpaired electron in an outer orbital... These radicals are highly reactive owing to tendency of the radical to either add an electron to the unpaired one or to donate the extra electron. Both these processes result in generation of another free radical in the bargain. Free radical generation is a continuous process in the body as a result of oxireductive processes in the mitochondria where four electrons are added to each oxygen molecule and the product is two molecules of water... This is the end result of a four-stage reaction, with an electron being added in each stage, producing superoxide ($O2^-$), hydrogen peroxides (H_2O_2), hydroxyl (OH) and finally water (H_2O)." See also Pallavi Sharma et al., "Reactive Oxygen Species, Oxidative Damage, and Antioxidative Defense Mechanism in Plants under Stressful Conditions," *Journal of Botany* (April 24, 2012): 1–26, https://doi.org/10.1155/2012/217037.

68. Andrew L. Rose, James W. Moffett, and T. David Waite, "Determination of Superoxfree Radicals in Seawater Using 2-Methyl-6-(4-Methoxyphenyl)-3,7- Dihydroimidazo[1,2-a] Pyrazin-3(7 H)-One Chemiluminescence," *Analytical Chemistry* 80, no. 4 (February 1, 2008): 1215–1227. https://doi.org/10.1021/ac7018975; Tong Zhang et al., "Extensive Dark Biological Production of Reactive Oxygen Species in Brackish and Freshwater Ponds," *Environmental Science & Technology* 50, no. 6 (March 15, 2016): 2983–2993, https://doi.org/10.1021/acs.est.5b03906.

69. T. P. A. Devasagayam et al., "Free Radicals and Antioxidants."

70. Fridovich, "Oxygen Toxicity."

71. Lien Pham-Huy, Ai Hua He, and Chuong Pham-Huy, "Free Radicals, Antioxidants in Disease and Health," *International Journal of Biomedical Science* 4, no. 2 (June 2008): 89–96; G. Pizzino et al., "Oxidative Stress: Harms and Benefits for Human Health," *Oxidative Medicine and Cellular Longevity* 2017 (2017): 8416763, https://doi.org/10.1155/2017/8416763. Pizzino and his colleagues comment, "When maintained at low or moderate concentrations, free radicals play several beneficial roles for the organism. For example, they are needed to synthesize some cellular structures and to be used by

the host defense system to fight pathogens. In fact, phagocytes synthesize and store free radicals, in order to be able to release them when invading pathogenic microbes have to be destroyed… The pivotal role of ROS [reactive oxygen species] for the immune system is well exemplified by patients with granulomatous disease. These individuals are unable to produce $O_2^{\cdot-}$ because of a defective NADPH oxidase system, so they are prone to multiple and in most of the cases persistent infections… Free radicals are also involved in a number of cellular signalling pathways… They can be produced by nonphagocytic NADPH oxidase isoforms; in this case, free radicals play a key regulatory role in intracellular signaling cascades, in several cell types such as fibroblasts, endothelial cells, vascular smooth muscle cells, cardiac myocytes, and thyroid tissue. Probably, the most well-known free radical acting as a signaling molecule is nitric oxide (NO). It is an important cell-to-cell messenger required for a proper blood flow modulation, involved in thrombosis, and is crucial for the normal neural activity …NO is also involved in nonspecific host defense, required to eliminate intracellular pathogens and tumor cells. Another physiological activity of free radicals is the induction of a mitogenic response… Summarizing, free radicals, when maintained at low or moderate levels, are of crucial importance to human health." For a popular article on the subject, see "Free Radicals May Actually Be Good for Us," NHS, May 27, 2014, https://web.archive.org/web/20201111190623/https://www.nhs.uk/news/food-and-diet/free-radicals-may-actually-be-good-for-us/; and "Free Radicals May Be Good for You," *Science Daily*, March 1, 2011, https://www.sciencedaily.com/releases/2011/02/110228090404.htm.

72. E. E. R. Philipp and D. Abele, "Masters of Longevity: Lessons from Long-Lived Bivalves—a Mini-Review," *Gerontology* 56 (2010): 55–65, https://doi.org/10.1159/000221004.

73. R. M. Lanner, *The Bristlecone Book: A Natural History of the World's Oldest Trees* (Missoula, MT: Mountain Press, 2007).

74. T. C. J. Tan et al., "Telomere Maintenance and Telomerase Activity are Differentially Regulated in Asexual and Sexual Worms," *PNAS USA* 109 (2012): 4209–4214, https://doi.org/10.1073/pnas.1118885109. For the longevity of the jellyfish Turritopsis, see S. F. Gilbert, *Developmental Biology*, 9th ed. (Sunderland, MA: Sinauer Associates, 2010), chap. 2.

75. Feher, "Gas Exchange in the Lungs."

76. Irving, *The Physics of the Human Body*.

77. Irving, *The Physics of the Human Body*, 21 table 1.11. The table is available online at Bionumbers, https://bionumbers.hms.harvard.edu/bionumber.aspx?s=n&v=2&id=109719.

78. Emilia Huerta-Sánchez et al., "Altitude Adaptation in Tibetans Caused by Introgression of Denisovan-like DNA," *Nature* 512, no. 7513 (July 2, 2014): 194–197, https://doi.org/10.1038/nature13408; Tianyl Wu and Bengt Kayser, "High Altitude Adaptation in Tibetans," *High Altitude Medicine & Biology* 7, no. 3 (2006): 193–208, https://doi.org/10.1089/ham.2006.7.193; J. L. Rupert and P. W. Hochachka, "Genetic Approaches to Understanding Human Adaptation to Altitude in the Andes," *The Journal of Experimental Biology* 204, no. 18 (September 2001): 3151–3160; P. D. Wagner, "The Biology of Oxygen," *European Respiratory Journal* 31, no. 4 (April 1, 2008): 887–890, https://doi.org/10.1183/09031936.00155407.

79. Johan Petersson and Robb W. Glenny, "Gas Exchange and Ventilation-Perfusion Relationships in the Lung," *European Respiratory Journal* 44, no. 4 (October 2014): 1023–1041, https://doi.org/10.1183/09031936.00037014.

80. Irving, *The Physics of the Human Body*.

81. Irving, *The Physics of the Human Body*.

82. Catling et al., "Why O_2 Is Required by Complex Life on Habitable Planets."

83. See Karl S. Kruszelnicki, "Why the Sky is Blue," ABC Science, May 31, 2005, https://www.abc.net.au/science/articles/2005/05/31/1360804.htm.

6. CIRCULATION

1. William Harvey, *Exercitatio Anatomica de Motu Cordis et Sanguinis in Animalibus De Motu Cordis* [1628], trans. Chauncey D. Leake (Spingfield, IL: Thomas, 1928), 71.

2. Knut Schmidt-Nielsen, *Animal Physiology: Adaptation and Environment*, 5th ed. (Cambridge, UK: Cambridge University Press, 1997). See Appendix B, 585.

3. Schmidt-Nielsen, *Animal Physiology*, chap. 3.

4. For more on blood vessels and our cardiovascular system, see "Blood Vessels," Atlas of Human Cardiac Anatomy, University of Minnesota, http://www.vhlab.umn.edu/atlas/physiology-tutorial/blood-vessels.shtml, and the accompanying pages at that site on other aspects of the cardiovascular system.

5. Y. Marcus and G.T. Hefter, "The Compressibility of Liquids at Ambient Temperature and Pressure," *Journal of Molecular Liquids* 73–74 (November 1997): 61–74, https://doi.org/10.1016/S0167-7322(97)00057-3; Steven Vogel, *Comparative Biomechanics: Life's Physical World*, 2nd ed. (Princeton, NJ: Princeton University Press, 2013), 12; Ray E. Bolz and George L. Tuve, eds., *Handbook of Tables for Applied Engineering Science* (Cleveland, OH: CRC Press, 1973), 95, Table 1-50, "Isothermal Compressibility of Liquids."

6. Jha Alok, *The Water Book* (London: Headline, 2015), 24.

7. The only polar (or charged molecules) which do not dissolve are very large molecules like cotton or cellulose.

8. C. R. Nave, "Henry's Law," Hyperphysics, Georgia State University, http://hyperphysics.phy-astr.gsu.edu/hbase/Kinetic/henry.html.

9. Lawrence J. Henderson, *The Fitness of the Environment: An Enquiry into the Biological Significance of the Properties of Matter* (New York: McMillan, 1913), 137–139.

10. Glenn Elert, "Density," The Physics Hypertextbook, accessed March 13, 2022, http://physics.info/density/.

11. Vogel, *Comparative Biomechanics*, 186. Vogel states that the heart uses up 10.7 percent of resting metabolic energy.

12. Vogel, *Comparative Biomechanics*, chap. 3 section headed "Viscosity."

13. "Effect of Moisture Content on the Viscosity of Honey at Different Temperatures," *Journal of Food Processing* 72, no. 4 (February 2006), 372–377.

14. Claudio Peri, ed., *The Extra-Virgin Olive Oil Handbook* (Chichester, UK: John Wiley & Sons, 2014), 352.

15. Schmidt-Nielsen, *Animal Physiology*, 146.

16. This calculation assumes a capillary diameter of 7 microns and a figure of 3,000 capillaries per sq mm of muscle, given by Schmidt-Nielsen in *Animal Physiology*, chap. 3.

17. Schmidt-Nielsen, *Animal Physiology*, chap. 2. Capillary density in human muscles is about 1,000 capillaries per sq mm. See B. J. McGuire and T. W. Secomb, "Estimation of Capillary Density in Human Skeletal Muscle Based on Maximal Oxygen Consumption Rates," *American Journal of Physiology—Heart and Circulatory Physiology* 285, no. 6 (December 2003): H2382–2391, https://doi.org/10.1152/ajpheart.00559.2003.

18. Vogel, *Comparative Biomechanics*, chap. 10.

19. Michael J. Denton, "The Place of Life and Man in Nature: Defending the Anthropocentric Thesis," *BIO-Complexity* 2013, no. 1 (February 25, 2013): 7, https://doi.org/10.5048/bio-c.2013.1. Internal references removed.

20. Jia Dongdong et al., "The Time, Size, Viscosity, and Temperature Dependence of the Brownian Motion of Polystyrene Microspheres," *American Journal of Physics* 75, no. 2 (February 2007): 111–115, https://doi.org/10.1119/1.2386163.

21. For the requirement for a liquid matrix, see N. V. Sidgwick, "Molecules," *Science* 86 (1937): 335–340; J. A. Baross et al., *The Limits of Organic Life in Planetary Systems* (Washington, DC: National Academies Press, 2007), 1; Kevin W. Plaxco and Michael Gross, *Astrobiology: A Brief Introduction*, 2nd ed., (Baltimore: Johns Hopkins University Press, 2011), 14; Arthur Needham, *The Uniqueness of Biological Materials* (Oxford: Pergamon Press, 1965), 9; and Louis Irwin, *Cosmic Biology: How Life Could Evolve on Other Worlds* (New York: Springer and Praxis, 2011), 32, 43, 303. The latter author, for example, finds the necessity for a liquid matrix "compelling" and insists that liquids provide "overwhelming advantages" over solids and gases to serve as the matrix of life (32).

22. Needham, *Uniqueness*, 9.

23. Richard E. Klabunde, "Turbulent Flow," January 3, 2018, Cardiovascular Physiology Concepts, https://www.cvphysiology.com/Hemodynamics/H007; H. N. Sabbah and P. D. Stein, "Turbulent Blood Flow in Humans: Its Primary Role in the Production of Ejection Murmurs," *Circulation Research* 38, no. 6 (June 1976): 513–525, https://doi.org/10.1161/01.res.38.6.513.dynamics/H007.htm.

24. Carol Porth, ed., *Essentials of Pathophysiology: Concepts of Altered Health States* (Philadelphia: Wolters Kluwer/Lippincott Williams & Wilkins, 2011), 380, 381, 416.

25. Klabunde, "Turbulent Flow."

26. Paul Clements and Carl Gwinnutt, "The Physics of Flow," AneasthesiaUK, accessed March 13, 2022, http://www.frca.co.uk/Documents/100308%20Physics%20of%20flowLR.pdf.

27. Vogel, *Comparative Biomechanics*, see discussion in chap. 10, 189–193, of Murray's law and the relationship between shear stress and remodeling of arterial walls. See also T. F. Sherman, "On Connecting Large Vessels to Small: The Meaning of Murray's law," *Journal of General Physiology* 78 (October 1981): 431–453, https://doi.org/10.1085/jgp.78.4.431.

28. Yaolin Shi and Jianling Cao, "Lithosphere Effective Viscosity of Continental China," *Earth Science Frontiers* 15, no. 3 (May 2008): 82–95, https://doi.org/10.1016/S1872-5791(08)60064-0.

29. Glenn Elert, "Viscosity," The Physics Hypertextbook, accessed March 11, 2022, https://physics.info/viscosity/.

30. Elert, "Viscosity."

31. Andrew D. Wong et al., "The Blood-Brain Barrier: An Engineering Perspective," *Frontiers in Neuroengineering* 6 (2013), https://doi.org/10.3389/fneng.2013.00007. The authors comment, "From transmission electron microscope images of rat brain capillaries, the endothelial cell thickness ranges from about 0.2 μm away from the nucleus to about 0.9 μm in the vicinity of the nucleus."

32. For more on Laplace's law, see R. Nave, Hyperphysics, Georgia State University, accessed March 12, 2022, http://hyperphysics.phy-astr.gsu.edu/hbase/ptens.html.

33. Vogel, *Comparative Biomechanics*, 194.

34. Vogel, *Comparative Biomechanics*, chap. 10.

230 / Endnotes 7. Warm-Blooded

35. Cecil D. Murray, "The Physiological Principle of Minimum Work, I: The Vascular System and the Cost of Blood Volume," *PNAS* 12 (1926): 207–214, https://doi.org/10.1073/pnas.12.3.207.

36. For Laplace's law see Schmidt-Nielsen, *Animal Physiology*, 110, and Vogel, *Comparative Biomechanics*, 23.

37. For a more detailed discussion and review of this area, see Vogel's *Comparative Biomechanics*, chap. 10 and Schmidt-Nielsen's *Animal Physiology*, chap. 3.

38. Murray, "The Physiological Principle of Minimum Work," 208.

39. Vogel, *Comparative Biomechanics*, 187.

40. S. L. Lindstedt and P. J. Schaeffer, "Use of Allometry in Predicting Anatomical and Physiological Parameters of Mammals," *Laboratory Animals* 36, no. 1 (January 1, 2002): 10, Figure 5, https://doi.org/10.1258/0023677021911731. The blood takes up an increased proportion of the volume of the body in diving mammals where it acts as a reservoir of oxygen. See Knut Schmidt-Nielsen, *Scaling: Why is Animal Size so Important?* (Cambridge, UK: Cambridge University Press, 1984), chap. 10, section headed "Blood Volume."

41. Vogel, *Comparative Biomechanics*, 187.

42. Schmidt-Nielsen, *Scaling*, chap. 10, section "Blood Volume."

43. Schmidt-Nielsen, *Animal Physiology*, chap. 3, Figure 3.9.

44. Craig R. White and Roger S. Seymour, "The Role of Gravity in the Evolution of Mammalian Blood Pressure: Brief Communication," *Evolution* 68, no. 3 (March 2014): 901–908, https://doi.org/10.1111/evo.12298.

45. Human red blood cells are about 8 microns in diameter. See Monica Diez-Silva et al., "Shape and Biomechanical Characteristics of Human Red Blood Cells in Health and Disease," *MRS Bulletin* 35, no. 5 (May 2010): 382–388, https://www.ncbi.nlm.nih.gov/pmc/articles/PMC2998922/. The various kinds of white blood cells range from about 7–30 microns in diameter. See "White Blood Cell," Wikimedia Foundation, last modified March 8, 2022, https://en.wikipedia.org/wiki/White_blood_cell.

46. C. M. Doerschuk, "Neutrophil Rheology and Transit through Capillaries and Sinusoids," *American Journal of Respiratory and Critical Care Medicine* 159, no. 6 (June 1999): 1693–1695, https://doi.org/10.1164/ajrccm.159.6.ed08-99. The authors comment, "Neutrophils... travel in hops through the highly anastomosing network of pulmonary capillary segments, with pauses followed by rapid travel. These observations suggest that a neutrophil moves rapidly through the larger capillary segments, but when it encounters a narrow one, it stops and requires time to deform. While neutrophils appear to elongate quickly when only small changes in shape are needed to enter a segment, deformation for entry into segments with diameters less than about 5.3 μm requires a delay." See also Jens G. Danielczok et al., "Red Blood Cell Passage of Small Capillaries Is Associated with Transient Ca2+-Mediated Adaptations," *Frontiers in Physiology* 8 (December 5, 2017): 979, https://doi.org/10.3389/fphys.2017.00979.

47. P. H. Irving, *The Physics of the Human Body: A Physical View of Physiology Biological and Medical Physics, Biomedical Engineering* (New York: Springer, 2007), 21, table 1.11.

48. Irving, *The Physics of the Human Body*, 21.

7. WARM-BLOODED

1. Lawrence J. Henderson, *The Fitness of the Environment: An Enquiry into the Biological Significance of the Properties of Matter* (New York: McMillan, 1913), vii.

2. William Hale-White, "Theory to Explain the Evolution of Warm-Blooded Vertebrates," *Journal of Anatomy and Physiology* 25, Fourth Part (April 1891): 374–385. Tom Garrison emphasized the same point more than a century later. "Remember the hot sand and cool water on a hot summer afternoon? The highest temperatures on land in the north African deserts exceed 50C… the lowest on the Antarctic continent, drop below −90C… That's a difference of 140C… On the ocean surface, however, the range is from −2C… where sea ice is forming to about 32 degrees in the tropics, a difference of only 34C… the ocean's thermal inertia is much greater than the land's." *Oceanography: An Invitation to Marine Science*, 7th ed. (Belmont, CA: Cengage Learning, 2010), 163.

3. Flying insects such as bees and dragonflies are endothermic and thermoregulate during flight as their flight muscles heat up to temperatures between 35–40°C, which may be well above their surroundings, especially in the higher latitudes. Shivering is also carried out in the nests of social insects to raise the temperature of the nest to temperatures well above that of the surrounding air temperature. See Bernd Heinrich, *Bumblebee Economics* (Cambridge, MA: Harvard University Press, 2004), chaps. 3 and 4; and Randy Oliver, "Old Bees? Cold Bees? No Bees?," Scientific Beekeeping, July 2008, http://scientificbeekeeping.com/old-bees-cold-bees-no-bees-part-2/.

4. Geoffrey K. Vallis, *Climate and the Oceans*, Princeton Primers in Climate (Princeton, NJ: Princeton University Press, 2012), see chap. 7.

5. Michael J. Angilletta, *Thermal Adaptation: A Theoretical and Empirical Synthesis* (Oxford, UK: Oxford University Press, 2009), 96.

6. The tradeoff is that because our bodily functions are all adapted to work at or very near a particular temperature (37°C/98.6°F in humans), significant departures from the normal body temperature in endotherms can be life threatening. Only in special cases, such as in hibernation, do endotherms survive a drastic change in body temperature—bears in the North American winter, for example—and only because the organisms possess special adaptations to allow periodic lowering of metabolic rates and body temperatures.

7. Angilletta, *Thermal Adaptation*, 95, and see section 4.2.

8. Heinrich, *Bumblebee Economics*.

9. A. F. Bennett, "Temperature and Muscle," *The Journal of Experimental Biology* 115 (March 1985): 333–344; A. F. Bennett, "Thermal Dependence of Muscle Function," *The American Journal of Physiology* 247, no. 2 part 2 (August 1984): R217–229.

10. David C. Preston and Barbara Ellen Shapiro, *Electromyography and Neuromuscular Disorders: Clinical-Electrophysiologic Correlations*, 3rd ed. (London: Elsevier Saunders, 2013), 327, 328, figure 13.1.

11. Lauren Augustine, "The Croc and Gator Blog," Smithsonian's National Zoo & Conservation Biology Institute, July 18, 2015, https://nationalzoo.si.edu/animals/news/croc-and-gator-blog-jun-18-2015.

12. C. D. Bramwell and P. B. Fellgett, "Thermal Regulation in Sail Lizards," *Nature* 242 (1973): 203–205, https://doi.org/10.1038/242203a0.

13. Kerstin A. Fritsches, Richard W. Brill, and Eric J. Warrant, "Warm Eyes Provide Superior Vision in Swordfishes," *Current Biology* 15, no. 1 (January 2005): 55–58, https://doi.org/10.1016/j.cub.2004.12.064.

14. Fritsches, Brill, and Warrant, "Warm Eyes." Also, research fish biologist Nicholas Wegner commented on the advantages of warm-bloodedness in an article describing the newly discovered first endothermic fish, the opha. Due to its warm-bloodedness, "the muscles can contract faster, the temporal resolution of the eye is increased, and neurological transmis-

sions are sped up," Wegner said. "This results in faster swimming speeds, better vision and faster response times." Quoted in Stefanie Pappas, "First Warm-Blooded Fish Discovered," *Scientific American* (May 14, 2015), https://www.scientificamerican.com/article/first-warm-blooded-fish-discovered/.

15. For more on the cleverness of some water-breathing sea-goers, see F. B. M. de Waal, *Are We Smart Enough to Know How Smart Animals Are?* (New York: W. W. Norton & Company, 2016); Gene S. Helfman, Bruce B. Collette, and Douglas E. Facey, *The Diversity of Fishes: Biology, Evolution, and Ecology*, 2nd ed. (Chichester, UK: Blackwell, 2009), 263; Peter Godfrey-Smith, *Other Minds: The Octopus, the Sea, and the Deep Origins of Consciousness* (New York: Farrar, Straus and Giroux, 2016); Damian Carrington, "Blue Planet 2: Attenborough Defends Shots Filmed in Studio," *Guardian*, October 23, 2017, https://www.theguardian.com/environment/2017/oct/23/captive-wildlife-footage-blue-planet-2-bbc1-totally-true-to-nature-say-producers.

16. Fritsches, Brill, and Warrant, "Warm Eyes." See also Ursula Dicke and Gerhard Roth, "Neuronal Factors Determining High Intelligence," *Philosophical Transactions of the Royal Society B: Biological Sciences* 371, no. 1685 (January 5, 2016): 20150180, https://doi.org/10.1098/rstb.2015.0180. As the authors note, axonal conduction velocity is an important determinant of a brain's information-processing capacity (or general intelligence).

17. David C. Catling et al., "Why O_2 Is Required by Complex Life on Habitable Planets and the Concept of Planetary 'Oxygenation Time,'" *Astrobiology* 5, no. 3 (June 2005): 415–438, https://doi.org/10.1089/ast.2005.5.415.

18. A. E. Needham, *The Uniqueness of Biological Materials* (London: Pergamon Press, 1965) 13; Henderson, *Fitness*, 81. And see Martin Chaplin, "Explanation of the Thermodynamic Anomalies of Water (T1-T11)," Water Structure and Science, October 30, 2021, https://water.lsbu.ac.uk/water/thermodynamic_anomalies.html. "Water has the highest specific heat of all liquids except ammonia," Chaplin writes. "The values for C_V and C_P are 4.1375 J g^{-1} K^{-1} and 4.1819 J g^{-1} K^{-1} at 25°C respectively (compare C_P pentane 1.66 J g^{-1} K^{-1}). As water is heated, the increased movement of water causes the hydrogen bonds to bend and break. As the energy absorbed in these processes is not available to increase the kinetic energy of the water, it takes considerable heat to raise water's temperature. Also, as water is a light molecule there are more molecules per gram, than most similar molecules, to absorb this energy. Heat absorbed is given out on cooling, so allowing water to act as a heat reservoir, buffering against changes in temperature. Thus, the water in our oceans stores vast amounts of energy, so moderating Earth's climate."

19. Garrison, *Oceanography*, 164.

20. Henderson, *Fitness*, 89.

21. Henderson, *Fitness*, 90.

22. For a brief video of a dragonfly shivering to warm up, see "Shivering to Warm Flight Muscles," August 26, 2013, YouTube, video, 0:29, https://www.youtube.com/watch?v=8Ta-O8cEoTQ.

23. Chaplin, "Explanation of the Thermodynamic Anomalies of Water." Chaplin notes, "The (isobaric; also called isopiestic) specific heat capacity (C_P) has a shallow minimum at about 36°C at 100 kPa… It is interesting that this minimum is close to the body temperature of warm-blooded animals."

24. Needham, *Uniqueness*, 10.

25. "Tissue Properties: Database," ItIs Foundation, https://itis.swiss/virtual-population/tissue-properties/database/heat-capacity/.

26. Knut Schmidt-Nielsen, *Animal Physiology: Adaptation and Environment*, 5th ed. (Cambridge, UK: Cambridge University Press, 1997), 252.

27. Henderson, *Fitness*, 98.

28. Needham, *Uniqueness*, 13.

29. "Heat of Fusion and Vaporization," Chemistry 301, University of Texas, 2014, https://ch301.cm.utexas.edu/data/section2.php?target=heat-transition.php.

30. Chaplin, "Explanation of the Thermodynamic Anomalies of Water." Chaplin writes, "Water has the highest heat of vaporization per gram of any molecular liquid (2257 J x g^{-1} at boiling point)... There is still considerable hydrogen bonding (~75%) in water at 100°C. As effectively all these bonds need to be broken (very few indeed remaining in the gas phase), there is a great deal of energy required to convert the water to gas, where the water molecules are effectively separated. The increased hydrogen bonding at lower temperatures causes higher heats of vaporization (for example, 44.8 kJ mol^{-1}, at 0°C)." See also table 7 in Needham, *Uniqueness*, 14.

31. Henderson, *Fitness*, 102–103. See also Schmidt-Nielsen, *Animal Physiology*, chap. 7, 271, figure 7.21. Although most significant at temperatures of 37°C or above, evaporative cooling of water is still an important means of heat loss at temperatures well below body temperature.

32. A good account of the physics of heat transfer and temperature regulation, including the roles of radiation and conduction as applied to biological systems, is given in chap. 7 of Schmidt-Nielsen's *Animal Physiology*.

33. D. Lieberman et al., "Brains, Brawn, and the Evolution of Human Endurance Running Capabilities," in *The First Humans: Origin and Early Evolution of the Genus Homo, Contributions from the Third Stony Brook Human Evolution Symposium and Workshop, October 3–7, 2006*, eds. F. E. Grine, John G. Fleagle, and Richard E. Leakey (Dordrecht: Springer, 2009), 85.

34. Michael Sockol, David Raichlen, and Herman Pontzer, "Chimpanzee Locomotor Energetics and the Origin of Human Bipedalism," *Proceedings of the National Academy of Sciences* 104, no. 30 (July 24, 2007): 12265–12269, https://doi.org/10.1073/pnas.0703267104.

35. Sockol, Raichlen, and Pontzer, "Chimpanzee Locomotor Energetics."

36. "Anhidrosis," Mayo Clinic, accessed March 14, 2022, http://www.mayoclinic.org/diseases-conditions/anhidrosis/basics/definition/con-20033498.

37. Colin Raymond, Tom Matthews, and Radley M. Horton, "The Emergence of Heat and Humidity Too Severe for Human Tolerance," *Science Advances* 6, no. 19 (May 2020), https://doi.org/10.1126/sciadv.aaw1838. Internal references removed.

38. Henderson, *Fitness*, 106.

39. Schmidt-Nielsen, *Animal Physiology*, chap. 7, 249, table 7.3.

40. Schmidt-Nielsen, *Animal Physiology*, chap. 7, 249, table 7.3.

41. Chaplin, "Explanation of the Thermodynamic Anomalies of Water." Chaplin writes, "Apart from liquid metals, water has the highest thermal conductivity of any liquid. For most liquids the thermal conductivity (the rate at which energy is transferred down a temperature gradient) falls with increasing temperature but this occurs only above about 130°C in liquid water." See also Needham, *Uniqueness*, 22; and "Solids, Liquids and Gases—Thermal Conductivities," The Engineering ToolBox, 2003, http://www.engineeringtoolbox.com/thermal-conductivity-d_429.html. Out of fifty-five common liquids cited, only mercury has a higher heat conductivity than water.

42. Since conduction works much too slowly over longer distances, another process is needed to conduct heat the much greater distance from the body's core to the periphery. That process is convection via blood circulation. As we saw in a previous chapter, here too water is an ideal medium for such a process. In the case of convection, in contrast to conduction, heat is absorbed by some substance, generally a gas or fluid, which is then moved so as to carry the heat from the absorption source to another place where the heat is given off. Unlike conduction, convection efficiently transports heat rapidly over long distances. Heat transference by convection is used in a blow heater or a fan to redistribute or circulate heat within a room. Central heating systems that pump heated water through pipes also exploit convection to redistribute heat, in the same manner as the circulatory system in the body.

43. Chaplin, "Explanation of the Thermodynamic Anomalies of Water."

44. See "Water—Thermal Conductivity vs. Temperature," The Engineering ToolBox, 2003, https://www.engineeringtoolbox.com/water-liquid-gas-thermal-conductivity-temperature-pressure-d_2012.html; and "Air—Thermal Conductivity vs. Temperature and Pressure," The Engineering ToolBox, 2003, https://www.engineeringtoolbox.com/air-properties-viscosity-conductivity-heat-capacity-d_1509.html.

8. Oxygen: Delving Deeper

1. John Donne, "An Anatomy of the World" (1611), https://www.bartleby.com/357/172.html#241.

2. Lawrence J. Henderson, *The Fitness of the Environment: An Enquiry into the Biological Significance of the Properties of Matter* (New York: Macmillan, 2013), 240–241.

3. Robert J. P. Williams, "The Symbiosis of Metal and Protein Function," *European Journal of Biochemistry* 150 (1985): 246.

4. Kasper P. Jensen and Ulf Ryde, "How O_2 Binds to Heme: Reasons for Rapid Binding and Spin Inversion," *Journal of Biological Chemistry* 279, no. 15 (April 9, 2004): 14561–14569, https://doi.org/10.1074/jbc.M314007200.

5. David C. Catling et al., "Why O_2 Is Required by Complex Life on Habitable Planets and the Concept of Planetary 'Oxygenation Time,'" *Astrobiology* 5, no. 3 (June 2005): 415–438, https://doi.org/10.1089/ast.2005.5.415.

6. Nick Lane, *Oxygen: The Molecule That Made the World* (Oxford, UK: Oxford University Press, 2002), 120–121.

7. Xiongyi Huang and John T. Groves, "Oxygen Activation and Radical Transformations in Heme Proteins and Metalloporphyrins," *Chemical Reviews* 118, no. 5 (March 14, 2018): 2491–2553, https://doi.org/10.1021/acs.chemrev.7b00373; Corinna R. Hess, Richard W. D. Welford, and Judith P. Klinman, "Oxygen-Activating Enzymes, Chemistry of," in *Wiley Encyclopedia of Chemical Biology* (Hoboken, NJ: John Wiley & Sons, 2008), http://doi.wiley.com/10.1002/9780470048672.wecb431. The authors comment: "Nature has developed a diverse array of catalysts to overcome this kinetic barrier. These dioxygen-activating enzymes are divided into two classes: oxygenases and oxidases. Oxygenases incorporate directly at least one atom from dioxygen into the organic products of their reaction. Oxidases couple the reduction of dioxygen with the oxidation of substrate. Typically, enzymes that react with dioxygen contain transition metal ions and/or conjugated organic molecules as cofactors."

8. Catling et al., "Why O_2 Is Required by Complex Life."

9. Vienna University of Technology, "Switching Oxygen Molecules between a Reactive and Unreactive State," Phys.org, March 14, 2017, https://phys.org/news/2017-03-oxygen-molecules-reactive-unreactive-state.html.

10. Vienna University of Technology, "Switching Oxygen Molecules."

11. Vienna University of Technology, "Switching Oxygen Molecules."

12. Catling et al., "Why O_2 Is Required by Complex Life."

13. See Robert R. Crichton, *Biological Inorganic Chemistry: A New Introduction to Molecular Structure and Function*, 2nd ed. (Amsterdam: Elsevier, 2012); and Wolfgang Kaim, Brigitte Schwederski, and Axel Klein, *Bioinorganic Chemistry: Inorganic Elements in the Chemistry of Life: An Introduction and Guide*, 2nd ed. (Chichester, West Sussex, UK: Wiley, 2013).

14. E. Frieden, "Evolution of Metals as Essential Elements," in *Protein-Metal Interactions*, ed. M. Friedman (New York: Plenum Press, 1974), 22.

15. Robert Gennis and Shelagh Ferguson-Miller, "Structure of Cytochrome c Oxidase, Energy Generator of Aerobic Life," *Science* 269 (1995): 1063–1064, https://doi.org/10.1126/science.7652553. And see Bruce Alberts et al., *Molecular Biology of the Cell*, 4th ed. (New York: Garland Press, 2002), figure 14–27, https://www.ncbi.nlm.nih.gov/books/NBK26910/.

16. Antony Crofts, "Cytochrome Oxidase," University of Illinois at Urbana-Champaign (website), 1996, https://www.life.illinois.edu/crofts/bioph354/cyt_ox.html.

17. Darryl Horn and Antoni Barrientos, "Mitochondrial Copper Metabolism and Delivery to Cytochrome C Oxidase," *IUBMB Life* 60, no. 7 (July 2008): 421–429, https://doi.org/10.1002/iub.50.

18. Michael Denton, *The Miracle of the Cell* (Seattle, WA: Discovery Institute Press, 2020), 106.

19. Michael J. Denton, "The Place of Life and Man in Nature: Defending the Anthropocentric Thesis," *BIO-Complexity* 2013, no. 1 (February 25, 2013): 1–15, https://doi.org/10.5048/BIO-C.2013.1.

20. A. E. Needham, *The Uniqueness of Biological Materials* (London: Pergamon Press, 1965), 35.

21. R. T. Sanderson, *Chemical Bonds and Bonds Energy* (New York: Academic Press, 1971), 21.

22. Henderson, *Fitness*, 139–140.

23. G. J. Arthurs and M. Sudhakar, "Carbon Dioxide Transport," *Continuing Education in Anaesthesia Critical Care & Pain* 5, no. 6 (December 2005): 207–210, https://doi.org/10.1093/bjaceaccp/mki050. See also M. Lieberman and A. D. Marks, *Marks' Basic Medical Biochemistry: A Clinical Approach*, 3rd North American ed. (Philadelphia: Lippincott Williams & Wilkins, 2008), 50–51.

24. B. D. Rose, *Clinical Physiology of Acid - Base and Electrolyte Disorders* (New York: McGraw-Hill, 1977), 176.

25. James B. Claiborne, "Acid-Base Regulation," in *The Physiology of Fishes*, 2nd ed., eds. D. H. Evans and J. B. Claiborne (Boca Raton, FL: CRC Press, 1997), 182.

26. James B. Claiborne, S. L. Edwards, and A. I. Morrison-Shetlar, "Acid-Base Regulation in Fishes: Cellular and Molecular Mechanisms," *Journal of Experimental Zoology* 293 (2002): 302–319. See also S. W. Marshall and M. Grosell, "Ion Transport, Osmoregulation and Acid Base Balance," in *The Physiology of Fishes*, 3rd ed., eds. David H. Evans and James B. Claiborne (Boca Raton, FL: CRC Press, 2006), 177–230; and Martin Tresguerres and Trevor J. Hamilton, "Acid–Base Physiology, Neurobiology and Behaviour in Relation to CO_2-Induced Ocean Acidification," *The Journal of Experimental Biology* 220, no. 12 (June 15, 2017): 2136–2148, https://doi.org/10.1242/jeb.144113.

27. Claiborne, "Acid-Base Regulation."

28. Claiborne, "Acid-Base Regulation." Internal references removed.

29. Lieberman and Marks, *Marks' Basic Medical Biochemistry*.

30. The Henderson-Hasselbalch equation, described in all major physiological texts.

31. Henderson, *Fitness*, 153.

32. J. T. Edsall and J. Wyman, *Biophysical Chemistry, Vol. 1: Thermodynamics, Electrostatics, and the Biological Significance of the Properties of Matter* (New York: Academic Press, 1955), chap. 10. As they show, every detail of this buffer system reveals further aspects to its fitness. For example, take the actual process of hydration itself. They write, "The hydration of CO_2 to H_2CO_3 is a process requiring a rearrangement of the valence bond of the two C—O bonds of CO_2, 180° apart and 1.15 Å long, being transformed to the three C—O bonds of H_2CO_3, approximately 120° apart and not far from 1.3 Å long. We shall not attempt to comment here on the details of the electronic rearrangements that must be involved in the process, and indeed little is known of them. It is not surprising, however, that a process such as this should require an appreciable time, in contrast for example to a process such as the hydration of NH_3 to NH_4OH in which the hydration process simply involves the formation of a hydrogen bond between the unshared electron pair in the ammonia molecule as acceptor, and one of the hydrogens of a water molecule as donor" (554). This apparently esoteric point, the slowness of the hydration of CO_2, may be of considerable physiological importance. If hydration were instantaneous, this would mean that whenever CO_2 levels in the blood plasma or body tissues increased suddenly following some respiratory distress, this well might provoke a lethal acidosis. Note: The hydration of CO_2 and the formation of bicarbonate occurs in the red cells catalyzed by the enzyme carbonic anhydrase. But while the bicarbonate diffuses out of the red cell, the hydrogen ions are retained inside the cell (the red cell membrane is impermeable to hydrogen ions). Electric charge neutrality is maintained by what is called the "chloride shift," where for each negatively charged bicarbonate ion leaving the cell a negatively charged chloride ion enters the cell.

33. Denton, "The Place of Life and Man in Nature."

34. Fred Hoyle, "The Universe: Past and Present Reflections," *Engineering and Science* (November 1981): 8–12. There Hoyle writes, "A common sense interpretation of the facts suggests that a superintellect has monkeyed with physics, as well as with chemistry and biology, and that there are no blind forces worth speaking about in nature. The numbers one calculates from the facts seem to me so over-whelming as to put this conclusion almost beyond question" (12).

9. THE RIGHT PROPORTIONS

1. Abour H. Cherif et al., "Redesigning Human Body Systems: Effective Pedagogical Strategy for Promoting Active Learning and STEM Education," *Education Research International* 2012 (2012): 1, 2, https://doi.org/10.1155/2012/570404. Internal references removed. Internal quotation is from Ray Kurzweil, "Human Body Version 2.0," para. 5, KurzweilAI. net, presented at Time Magazine's "The Future of Life" Conference, February 2003, http://www.kurzweilai.net/human-body-version-20.

2. Steven Vogel, *Comparative Biomechanics: Life's Physical World*, 2nd ed. (Princeton: Princeton University Press, 2013), 187.

3. Ella Davies, "The World's Strongest Animal Can Lift Staggering Weights," BBC Earth, November 21, 2016, https://web.archive.org/web/20161124102437/http://www.bbc.com/earth/story/20161121-the-worlds-strongest-animal-can-lift-staggering-weights.

4. Knut Schmidt-Nielsen, *Scaling: Why is Animal Size So Important?* (Cambridge, UK: Cambridge University Press, 1984), 210–211.

5. Schmidt-Nielsen, *Scaling*. Even more impressive than an ant is the dung beetle, reportedly able to pull more than 1,000 times its weight, equivalent to a human pulling eight large African elephants. See Davies, "The World's Strongest Animals."

6. Rod Lakes, "Scaling Concepts," Biomechanics BME 315, University of Wisconsin, http://silver.neep.wisc.edu/~lakes/BME315ScalingStrength.html.

7. Michael Denton, *Nature's Destiny: How the Laws of Biology Reveal Purpose in the Universe* (New York: Free Press, 1998), 245.

8. Denton, *Nature's Destiny*, 246. See also Steven M. Block, "Nanometres and Piconewtons: the Macromolecular Mechanisms of Kinesin," *Trends in Cell Biology* 5 (1996): 169–175; and R. Anthony Crowther, Raúl Padron, and Roger Craig, "Arrangement of the Heads of Myosin in Relaxed Thick Filaments from Tarantular Muscle," *Journal of Molecular Biology* 184 (1985): 429–439.

9. Denton, *Nature's Destiny*, 246.

10. Knut Schmidt-Nielsen, *Animal Physiology: Adaptation and Environment*, 5th ed. (Cambridge, UK: Cambridge University Press, 1997), 414.

11. James D. Watson, *The Molecular Biology of the Gene*, 3rd ed. (Menlo Park, California: W. A. Benjamin, 1976), 98.

12. Watson, *The Molecular Biology of the Gene*, 86. Specific binding of a particular substrate (the key) to its binding site on the protein (the lock) depends on having a number of weak bonds arranged in very precise complementary spatial positions in both the substrate and the binding site. Decreasing the number of weak bonds to compensate for a hypothetical increase in their strength would not work, since specificity of binding would be lost. On the other hand, if the strength of weak bonds had been any weaker, it would have been impossible to fit a sufficient number on complementary surfaces to form a bond strong enough to withstand the hurly burly and continual thermal jiggling of the constituents within the cell. As things are, disruptive forces in the cell caused by the random collisions between particles are close to the forces exerted by the weak bonds.

13. Watson, *The Molecular Biology of the Gene*, 86.

14. Watson, *The Molecular Biology of the Gene*, 100. "This fact explains why enzymes can function so quickly, sometimes as often as 10^6 times per second. If enzymes were bound to their substrates by more powerful bonds (if weak bonds were stronger) they would act more slowly and beyond a certain strength enzymic action would be impossible."

15. Watson, *The Molecular Biology of the Gene*, 100.

16. Irving P. Herman, *Physics of the Human Body: Biological and Medical Physics, Biomedical Engineering* (New York: Springer, 2007), 21, table 1.11; Vogel, *Comparative Biomechanics*, 390.

17. Tim Taylor, "Nerves of Arm and Hand," Innerbody Research, July 16, 2019, https://www.innerbody.com/anatomy/nervous/arm-hand#continued.

18. Michael Denton, *Fire-Maker* (Seattle, WA: Discovery Institute Press, 2016), 58.

19. See S. T. Aw et al., "Three-Dimensional Vector Analysis of the Human Vestibuloocular Reflex in Response to High-Acceleration Head Rotations. II. Responses in Subjects with Unilateral Vestibular Loss and Selective Semicircular Canal Occlusion," *Journal of Neurophysiology* 76, no. 6 (December 1996): 4021–4030, https://doi.org/10.1152/jn.1996.76.6.4021. The authors further comment, "To achieve clear vision, signals from the semicircular canals are sent as directly as possible to the eye muscles: the connection

involves only three neurons, and is correspondingly called the *three neuron arc*. Using these direct connections, eye movements lag the head movements by less than 10 ms, and thus the vestibulo-ocular reflex is one of the fastest reflexes in the human body."

20. Jorik Nonnekes et al., "Mechanisms of Postural Instability in Hereditary Spastic Paraplegia," *Journal of Neurology* 260, no. 9 (September 2013): 2387–2395, https://doi.org/10.1007/s00415-013-7002-3. See also Tina M. Weatherby et al., "The Need for Speed II: Myelin in Calanoid Copepods," *Journal of Comparative Physiology A: Sensory, Neural, and Behavioral Physiology* 186, no. 4 (April 2000): 347–357.

21. Schmidt-Nielsen, *Animal Physiology*, chap. 11, table 11.4.

22. Schmidt-Nielsen, *Scaling*, 213.

23. Schmidt-Nielsen, *Animal Physiology*, chap. 11, table 11.13. See also Dominique Debanne et al., "Axon Physiology," *Physiological Reviews* 91, no. 2 (April 1, 2011): 555–602, https://doi.org/10.1152/phys- rev.00048.2009.

24. Schmidt-Nielsen, *Scaling*, 213.

25. William R. Holmes, "Cable Equation," *Encyclopedia of Computational Neuroscience*, March 31, 2014, https://doi.org/10.1007/978-1-4614-7320-6_478–1.

26. D. K. Hartline and D. R. Colman, "Rapid Conduction and the Evolution of Giant Axons and Myelinated Fibers," *Current Biology* 17, no. 1 (January 2007): R29–35, https://doi.org/10.1016/j.cub.2006.11.042; J. Z. Young, "The Functioning of the Giant Nerve Fibres of the Squid," *Journal of Experimental Biology* 15, no. 2 (April 1, 1938): 170–185, https://doi.org/10.1242/jeb.15.2.170.

27. D. C. Davies et al., "Myelinated Axon Number in the Optic Nerve is Unaffected by Alzheimer's Disease," *The British Journal of Ophthalmology* 79, no. 6 (June 1995): 596–600, https://bjo.bmj.com/content/79/6/596.short.

28. Schmidt-Nielsen, *Animal Physiology*, 485.

29. Schmidt-Nielsen, *Animal Physiology*, 483.

30. Robert R. Crichton, *Biological Inorganic Chemistry: A New Introduction to Molecular Structure and Function*, 2nd ed. (Amsterdam: Elsevier, 2012), 5. Also see 197, table 10.1, showing correlations between ligand binding, mobility, and function of some of the biologically relevant metals.

31. Bruce Alberts et al., *Molecular Biology of the Cell*, 4th ed. (New York: Garland Press, 2002), https://www.ncbi.nlm.nih.gov/books/NBK26910/.

32. Crichton, *Biological Inorganic Chemistry*, 5.

33. Ursula Dicke and Gerhard Roth, "Neuronal Factors Determining High Intelligence," *Philosophical Transactions of the Royal Society B: Biological Sciences* 371, no. 1685 (January 5, 2016): 20150180, https://doi.org/10.1098/rstb.2015.0180.

34. Dicke and Roth, "Neuronal Factors Determining High Intelligence."

35. Dicke and Roth, "Neuronal Factors Determining High Intelligence."

36. Dicke and Roth, "Neuronal Factors Determining High Intelligence."

37. Michel A. Hofman, "Evolution of the Human Brain: When Bigger Is Better," *Frontiers in Neuroanatomy* 8 (March 27, 2014), https://doi.org/10.3389/fnana.2014.00015.

38. Peter Cochrane, C. S. Winter, and A. Hardwick, "Biological Limits to Information Processing in the Human Brain," Peter Cochrane (website), last modified November 16, 2004, https://www.gwern.net/docs/iq/1995-cochrane-biologicallimitstoinformationprocessinginthebrain.html.

39. Michel A. Hofman, "On the Nature and Evolution of the Human Mind," *Progress in Brain Research* 250 (2019): 251–283, https://doi.org/10.1016/bs.pbr.2019.03.016.

40. Christof Koch, "Computation and the Single Neuron," *Nature* 385 (1997), 207–210, https://doi.org/10.1038/385207a0.

41. Dickie and Roth, "Neuronal Factors."

42. Christopher S. von Bartheld, Jami Bahney, and Suzana Herculano-Houzel, "The Search for True Numbers of Neurons and Glial Cells in the Human Brain: A Review of 150 Years of Cell Counting: Quantifying Neurons and Glia in Human Brain," *Journal of Comparative Neurology* 524, no. 18 (May 17, 2016): 3865–3895, https://doi.org/10.1002/cne.24040.

43. Sarah Jäkel and Leda Dimou, "Glial Cells and Their Function in the Adult Brain: A Journey through the History of Their Ablation," *Frontiers in Cellular Neuroscience* 11 (February 13, 2017), https://doi.org/10.3389/fncel.2017.00024. For function of glial cells, see Cai-Yun Liu et al., "Emerging Roles of Astrocytes in Neuro-Vascular Unit and the Tripartite Synapse With Emphasis on Reactive Gliosis in the Context of Alzheimer's Disease," *Frontiers in Cellular Neuroscience* 12 (July 10, 2018): 193, https://doi.org/10.3389/fncel.2018.00193.

44. Koch, "Computation and the Single Neuron."

45. Chet C. Sherwood et al., "Invariant Synapse Density and Neuronal Connectivity Scaling in Primate Neocortical Evolution," *Cerebral Cortex* 30, no. 10 (September 3, 2020): 5604–5615, https://doi.org/10.1093/cercor/bhaa149.

46. "Boeing 747 Specs," Modern Airliners, accessed March 14, 2022, https://modernairliners.com/boeing-747-jumbo/boeing-747-specs/.

47. The only class of vertebrates lacking bone are the cyclostomes as well as sharks and rays, which substitute cartilage for bone.

48. G. Donald Whedon and Robert Proulx Heaney, "Bone," Encyclopedia Britannica, March 19, 2019, https://www.britannica.com/science/bone-anatomy, accessed March 15, 2022, https://www.britannica.com/science/bone-anatomy. The bones contain approximately 99 percent of the body's calcium, 85 percent of its phosphate, and 50 percent of its magnesium. See Reiner Bartl and Christoph Bartl, *Bone Disorders: Biology, Diagnosis, Prevention, Therapy* (Switzerland: Springer, 2017), 11.

49. Whedon and Heaney, "Bone."

50. Bartl and Bartl, *Bone Disorders*, 6.

51. Whedon and Heaney, "Bone." Under the subheading "Chemical Composition and Physical Properties," they write, "Compact (cortical) bone specimens have been found to have tensile strength in the range of 700–1,400 kg per square cm (10,000–20,000 pounds per square inch) and compressive strengths in the range of 1,400–2,100 kg per square cm (20,000–30,000 pounds per square inch). These values are of the same general order as for aluminum or mild steel, but bone has an advantage over such materials in that it is considerably lighter. The great strength of bone exists principally along its long axis and is roughly parallel both to the collagen fibre axis and to the long axis of the mineral crystals."

52. Herman, *The Physics of the Human Body*, 21.

53. The fraction of the body comprised of bones varies far more across species than in the case of lungs, blood, and muscles (which, as mentioned previously, comprise approximately the same proportion in all mammalian species). As Galileo noted, the bones of large organisms are far larger as a proportion of total body volume than those in small organisms. Vogel explains: "The problem faced by the large creature is that a structure stressed by twisting, bending, crushing, or anything other than simple pulling (tension) is relatively weaker as it gets larger. In compensation, the scaling factor for skeletal mass is greater than 1.0—it's

1.08. Over the range of mammalian size, the difference is appreciable. An 8-gram shrew is about 4 percent skeleton; a 5-kilogram cat about 7; a 60-kilogram person about 8.5; a 600-kilogram horse 10; a 7000-kilogram elephant nearly 13 percent. From the proportions of the bones, you can guess from a drawing or photograph of its skeleton the size of an animal." Steven Vogel, *Comparative Biomechanics*, 559. See also "Power Law for Bone Mass of Extant Mammals," Bionumbers, accessed March 14, 2022, https://bionumbers.hms.harvard.edu/bionumber.aspx?s=n&v=2&id=108641.

54. For examples of sightless individuals managing extraordinary things, see Joanna Moorhead, "Seeing with Sound," *Guardian*, January 27, 2007, https://www.theguardian.com/lifeandstyle/2007/jan/27/familyandrelationships.family2; and "Humans With Amazing Senses," ABC News, March 14, 2008, https://abcnews.go.com/Primetime/story?id=2283048&page=1.

55. Kenneth C. Catania, "A Nose for Touch," *Scientist* (September 1, 2012), http://www.the-scientist.com/?articles.view/articleNo/32505/title/A-Nose-for-Touch/. Catania comments concerning the star-nosed mole, "In total, a single star contains about 25,000 domed Eimer's organs, each one served by four or so myelinated nerve fibers and probably about as many unmyelinated fibers. This adds up to many times more than the total number of touch fibers (17,000) found in the human hand—yet the entire star is smaller than a human fingertip." See also K. C. Catania and J. H. Kaas, "Areal and Callosal Connections in the Somatosensory Cortex of the Star-nosed Mole," *Somatosensory & Motor Research* 18 (2001): 303–311.

56. Carl Sagan, *Cosmos* (New York: Ballantine Books, 2013), 102.

57. One playing card is .25mm thick, the distance to the Andromeda galaxy is 2.5 million light years or 2.5×10^{25} mm. The visual region occupies a fraction smaller than one part in 10^{25} of the electromagnetic spectrum. (See discussion in Chapter 3.)

58. G. B. Airy, "On the Diffraction of an Object-glass with Circular Aperture," *Transactions of the Cambridge Philosophical Society*, 5 (1835): 283–291, https://archive.org/stream/transactionsofca05camb#page/n305/mode/2up/search/airy.

59. Gregory Hollows and Nicholas James, "The Airy Disk and Diffraction Limit," Edmund Optics, accessed March 13, 2022, https://www.edmundoptics.com/resources/application-notes/imaging/limitations-on-resolution-and-contrast-the-airy-disk/.

10. FIRE AND METAL

1. Stuart Ross Taylor and Scott M. McLennan, *Planetary Crusts: Their Composition, Origin and Evolution* (Cambridge, UK: Cambridge University Press, 2009), 1.

2. A. J. Wilson, *The Living Rock: The Story of Metals since the Earliest Times and Their Impact on Civilization* (Cambridge, UK: Woodhead Publishing Limited, 1994), 9.

3. Adam L. Penenberg, "Intel Atom: Intel Makes Its Smallest Chip Ever," *Fast Company*, October 1, 2008, http://www.fastcompany.com/1007035/intel-atom-intel-makes-its-smallest-chip-ever.

4. Michael Denton, *Fire-Maker: How Humans Were Designed to Harness Fire and Transform Our Planet*, Privileged Species Series (Seattle, WA: Discovery Institute Press, 2016), 10.

5. Richard Feynman, "There's Plenty of Room at the Bottom," nanotechnology lecture, December 29, 1969, Pasadena, CA, https://speakola.com/ideas/richard-feynman-nanotechnology-lecture-1959.

6. Eric Drexler, *The Engines of Creation* (New York: Anchor Press, 1990).

7. Denton, *Fire-Maker*, 9–10. Internal reference removed.

8. Charles Darwin, The *Descent of Man and Selection in Relationship to Sex*, vol. 1, 1st ed. (London: John Murray, 1871), 137.

9. J. A. J. Gowlett, "The Discovery of Fire by Humans: A Long and Convoluted Process," *Philosophical Transactions of the Royal Society B: Biological Sciences* 371, no. 1696 (June 5, 2016), https://doi.org/10.1098/rstb.2015.0164.

10. Gowlett, "The Discovery of Fire by Humans."

11. Gowlett, "The Discovery of Fire by Humans."

12. Gowlett, "The Discovery of Fire by Humans."

13. Xiaohong Wu et al., "Early Pottery at 20,000 Years Ago in Xianrendong Cave, China," *Science*, 336, no. 6089 (June 29, 2012), https://doi.org/10.1126/science.1218643.

14. Precisely how it was discovered that metals could be smelted from their ores is conjectural, but presumably this occurred by the chance finding of metal in the ashes of a hot fire that just happened to contain metal-bearing ore mixed with the wood or charcoal. See A. J. Wilson, *The Living Rock*, chap. 2; see also R. F. Tylecote, *A History of Metallurgy* (London: Maney Publication for the Institute of Materials, 2002), chap. 2.

15. Bruce Bower, "Serbian Site May Have Hosted First Copper Makers," ScienceNews, June 25, 2010, https://www.sciencenews.org/article/serbian-site-may-have-hosted-first-copper-makers.

16. Wilson, *The Living Rock*, 15, 52. See also Jay King, "The Emergence of Iron Smelting and Blacksmithing: 900 BC to the Early Roman Empire," Roman History, Coins, and Technology, 2006, http://www.jaysromanhistory.com/romeweb/glossary/timeln/t10.htm; Lee Horne, "Fuel for the Metal Worker," *Expedition Magazine* 25 no. 1 (1982), http://www.penn.museum/sites/expedition/?p=5281; and Peter Harris, "On Charcoal," *Interdisciplinary Science Reviews* 24, no. 4 (April 1999): 301–306, https://doi.org/10.1179/030801899678966.

17. Horne, "Fuel for The Metal Worker"; Harris, "On Charcoal"; King, "The Emergence of Iron Smelting."

18. Thomas J. Straka, "Charcoal as a Fuel in the Ironmaking and Smelting Industries," *Advances in Historical Studies* 6, no. 1 (2017): 56–64, https://doi.org/10.4236/ahs.2017.61004.

19. Straka, "Charcoal as a Fuel." See also Horne, "Fuel for the Firemaker"; Harris, "On Charcoal"; and King, "The Emergence of Fire Smelting."

20. Wilson, *The Living Rock*, 11.

21. Chris Butler, "FC8: The Birth of Metallurgy and Its Impact," The Flow of History: A Dynamic and Graphic Approach to Teaching History, 2007, https://web.archive.org/web/20071010171250/http://www.flowofhistory.com/units/pre/1/FC8.

22. Stephen J. Pyne, *Vestal Fire: An Environmental History, Told through Fire, of Europe and Europe's Encounter with the World* (Seattle, WA: University of Washington Press, 1997), 42–43.

23. Nick Lane, *Oxygen: The Molecule That Made the World* (Oxford, UK: Oxford University Press, 2002), chaps. 6 and 10; Huang Xiongyi and John T. Groves, "Oxygen Activation and Radical Transformations in Heme Proteins and Metalloporphyrins," *Chemical Reviews* 118, no. 5 (March 14, 2018): 2491–2553; R. Boulatov, "Understanding the Reaction That Powers This World: Biomimetic Studies of Respiratory O_2 Reduction by Cytochrome Oxidase," *Pure and Applied Chemistry* 76, no. 2 (2004): 303–319, https://doi.org/10.1351/pac200476020303; Irwin Fridovich, "Oxygen: How Do We Stand It?" *Medical Principles and Practice* 22, no. 2 (2013): 131–137, https://doi.org/10.1159/000339212; Corinna R. Hess, Richard W. D. Welford, and Judith P. Klinman, "Chemistry of Oxygen-Activating

Enzymes," *Wiley Encyclopedia of Chemical Biology* (Hoboken, NJ: John Wiley & Sons, Inc., 2008), http://doi.wiley.com/10.1002/9780470048672.wecb431.

24. See Michael Denton, *The Wonder of Water: Water's Profound Fitness for Life on Earth and Mankind* (Seattle, WA: Discovery Institute Press, 2017), chap. 5; and Vogel, *The Life of a Leaf* (Chicago, IL: University of Chicago Press, 2010), chap 6.

25. Vogel, *The Life of a Leaf,* 101.

26. Vogel, *The Life of a Leaf,* 99.

27. Horne, "Fuel for the Firemaker"; Straka, "Charcoal as a Fuel."

28. The extraction of metals from their ores requires not only heat but also reducing conditions to draw the oxygen from the metal. See Harris, "On Charcoal"; Horne, "Fuel for the Firemaker"; and King, "The Emergence of Fire Smelting." King comments: "The techniques developed by the copper workers did not generate enough heat to cause iron ore to give up its oxygen. Also, the quantities of carbon monoxide [which generates reducing condition in a kiln] had to be much greater than required for copper. Not only does iron melt at a higher temperature than copper, but iron oxide holds its oxygen atoms much more tenaciously than copper oxide does.... The answer was to first make a high quality charcoal from hardwoods.... The iron ore would be totally surrounded by charcoal and the furnace had to be more enclosed, having only a chimney to exhaust the fumes and inlets for the bellows supplying the air. Charcoal would first be loaded into the furnace, followed by the iron ore and more charcoal. The fires were lit, and the bellows operators pumped furiously to generate heat capable of adding enough energy to the iron oxide to make it loosen its grip on the oxygen atoms."

29. Wilson, *The Living Rock,* 11.

30. The tensile strength of an object, such as a wire, is the minimum pulling force required to break it.

31. Roy Beardmore, "Temperature Effects on Metal Strength," *RoyMech,* accessed on March 1, 2022, https://roymech.org/Useful_Tables/Matter/Temperature_effects.html.

32. Glenn Elert, "Electric Resistance," *The Physics Hypertextbook,* 1998–2015, accessed March 1, 2022, http://physics.info/electric-resistance.

33. "Copper Building Wire Systems," Copper Development Association Inc., accessed March 1, 2022, https://www.copper.org/applications/electrical/building/wire_systems.html.

34. Stuart Ross Taylor and Scott M McLennan, *Planetary Crusts: Their Composition, Origin and Evolution,* Cambridge Planetary Science (Cambridge, UK: Cambridge University Press, 2009), 7.

35. Marcia Bjornerud, *Reading the Rocks: The Autobiography of Earth* (New York: Basic Books, 2005), 119. She writes: "Lava viscosities vary hugely depending on the temperature and composition of the magma. A Hawaiian–type basaltic lava, extruded at more than... (1000 C) may have a viscosity of only 100 poise, while a Mount St. Helens rhyolite, a 'cold' lava at (800 C) may have a viscosity of 10 million poise. The difference is related mainly to the silica... content of the lava, and it has deadly implications. The extremely high viscosities of rhyolitic lavas inhibits the escape of magmatic gases and is responsible for Earth's most violent eruptions... [T]he earth's rocky mantle moves at the most languorous rate of all. Its viscosity has been estimated at 10^{21} poise... a trillion, trillion times that of water" (73–74.) See also chap. 2 of my book *The Wonder of Water.*

36. Sami Mikhail and Dimitri A. Sverjensky, "Nitrogen Speciation in Upper Mantle Fluids and the Origin of Earth's Nitrogen-Rich Atmosphere," *Nature Geoscience* 7 (October 19, 2014): 816.

37. "Rare Earth Elements—Critical Resources for High Technology," US Geological Survey, last modified May 17, 2005, http://pubs.usgs.gov/fs/2002/fs087-02/.

38. Michael Denton, *Children of Light: The Astonishing Properties of Sunlight That Make Us Possible* (Seattle, WA: Discovery Institute Press, 2018), 42.

11. THE FIRE MAKERS

1. Marcus Cicero, *On the Nature of the Gods*, trans. J. B. Mayor (London: Methuen), 45, https://oll.libertyfund.org/title/cicero-on-the-nature-of-the-gods.

2. Stephen Pyne, "The Ecology of Fire" *Nature Education Knowledge* 3, no. 10 (2010): 30.

3. F. B. M. de Waal, *Are We Smart Enough to Know How Smart Animals Are?* (New York: W. W. Norton & Company, 2016).

4. G. G. Simpson, *The Meaning of Evolution* (New Haven, CT: Yale University Press, 1967), 288.

5. Galen, *On the Usefulness of the Parts of the Body*, quoted in John Kidd, Bridgewater Treatise, *On the Adaptation of External Nature to the Physical Condition of Man* (London: H. G. Bohn, 1852), 26, 29.

6. Galen, *On the Usefulness of the Parts of the Body*, quoted in Kidd, *The Physical Condition of Man*, 33–34.

7. Charles Bell, Bridgewater Treatise, *The Hand: Its Mechanism and Vital Endowments as Evincing Design* (London: William Pickering 1834), 230–231, https://archive.org/details/handmechanismvit00belliala/page/230.

8. R. M. Yerkes and A. W. Yerkes, *The Great Apes* (New Haven, CT: Yale University Press, 1929), 346, https://archive.org/details/greatapesstudyof00yerk/page/346.

9. R. D. Martin, *Primate Origins and Evolution* (London: Chapman and Hall, 1990), 496–497.

10. Christopher Joyce, "A Handy Bunch: Tools, Thumbs Helped Us Thrive," NPR, July 26, 2010, https://www.npr.org/templates/story/story.php?storyId=128676181.

11. See Russell H. Tuttle, "Knuckle-Walking and Knuckle-Walkers: A Commentary on Some Recent Perspectives on Hominoid Evolution," in *Primate Functional Morphology and Evolution*, ed. R. Tuttle (The Hague, Netherlands: Mouton Publishers, 1975), 203.

12. Frits Warmolt Went, "The Size of Man," *American Scientist* 56, no. 4 (1968): 400–413, https://www.jstor.org/stable/27828330.

13. Hu Berry, "Ants Here, Flies There... Insects, 'Goggas' Everywhere!" NatureFriends Safaris, 2006, https://www.naturefriendsafaris.com/ants-here-flies-therinsects-goggas-everywhere/.

14. Michael J. Morwood et al., "Archaeology and Age of a New Hominin from Flores in Eastern Indonesia," *Nature* 431, no. 7012 (October 28, 2004): 1087–1091, https://doi.org/10.1038/nature02956.

15. Went, "The Size of Man."

16. Went, "The Size of Man." Note that because a 7-foot giant would be proportionately less strong than a 5 ft 8 in. man, his implements would not be exactly proportionate to his size.

17. Stephen Jay Gould, *The Richness of Life: The Essential Stephen Jay Gould* (New York: W. W. Norton, 2007), 321–322.

18. J. B. S. Haldane, "On Being the Right Size," *Harper's Magazine*, March 1926, available online at http://www.phys.ufl.edu/courses/phy3221/spring10/HaldaneRightSize.pdf.

19. Went, "The Size of Man." Steven Vogel comments, "The speed at which a large organism hits the ground when it trips is proportional to the square root of its height (drag in air be-

ing minimal for short falls by large creatures). Because its mass is proportional to the cube of its length (or height), the momentum mv with which it hits the ground is proportional to its length raised to the power 3.5, and the kinetic energy $\frac{1}{2}mv^2$ that has to be offloaded is proportional to its length to the fourth power." Steven Vogel, *Comparative Biomechanics: Life's Physical World*, 2nd ed. (Princeton, NJ: Princeton University Press, 2013), 556.

20. Vogel, *Comparative Biomechanics*, 18.

21. Vogel, *Comparative Biomechanics*, 556.

22. Some physicists think the inertial force experienced by objects is generated by the total combined gravitational attraction of all matter in the cosmos, including the most distant stars and galaxies. Because most matter in the universe is far from Earth, this would mean that the greatest contribution to the inertia of objects on Earth is made by very distant matter. Dennis Sciama comments in his *Unity of the Universe*: "The idea that distant matter can sometimes have far more influence than nearby matter may be an unfamiliar one. To make it more concrete, we may give a numerical estimate of the influence of nearby objects in determining the inertia of bodies on the earth: of this inertia, the whole of the Milky Way only contributes one ten millionth, the sun one hundred millionth, and the earth itself one thousand-millionth." As he continues at another point, "In fact, 80 per cent of the inertia of local matter arises from the influence of galaxies too distant to be detected by the 200 inch telescope." (D. W. Sciama, *The Unity of the Universe* (London: Faber and Faber, 1959), 181, 119.) If so, then the influence of very distant galaxies, whose collective mass largely determines the precise strength of the inertial forces on Earth, contributes indispensably to the existence of beings of our size and mass, ensuring our ability to stand, move, and start and control fires. This idea, which Einstein named Mach's Principle and which inspired Einstein in his work on formulating the general theory of relativity (See the Stanford Encyclopedia of Philosophy entry on Ernst Mach—https://plato.stanford.edu/entries/ernst-mach/), remains an open question in cosmology. There is in the idea and its implications for nature's fitness for us a distinct echo of the old medieval doctrine of man as microcosm, which held that the dimensions of the human body reflect in a profound sense the dimensions of the macrocosm. If true, it would be yet another intriguing piece of support for the anthropocentric thesis. It would mean that the cosmos that we now know to be unimaginably large, where Earth exists as a "pale blue dot" in an inconceivably vast ocean of stars and galaxies, turns out to be of vital necessity to the existence of beings such as ourselves.

23. T. H. Huxley, *Man's Place in Nature* (New York: Appleton and Company, 1863), 71.

24. A. J. Gurevich, *Categories of Medieval Culture* (London: Routledge and Kegan Paul, 1985), 57–61; Frederick Copleston, *A History of Philosophy*, vol. 3 (New York: Double Day, 1993), 242.

12. The End of the Matter

1. Francis Bacon, *The Advancement of Learning* [1605] (London: J. M Dent & Sons, 1915), 109, https://archive.org/details/advancementlearn00bacouoft/page/108.

2. Fred Hoyle, "The Universe Past and Present Reflections," *Engineering and Science* (November, 1981), 8–12.

3. "Encyclopaedia Galactica," *Cosmos: A Personal Voyage*, episode 12, PBS, written and hosted by Carl Sagan, aired December 14, 1980. See also Marcello Truzzi, "On the Extraordinary: An Attempt at Clarification," *Zetetic Scholar* 1, no. 1 (1978), 11.

4. Lawrence J. Henderson, *The Fitness of the Environment: An Enquiry into the Biological Significance of the Properties of Matter* (New York: McMillan, 1913), 312; S. L. Miller and L. E. Orgel, *The Origins of Life on the Earth* (Englewood Cliffs, NJ: Prentice-Hall, 1974); G.

Wald, "Fitness in the Universe: Choices and Necessities," *Origins Life* 5 (1974): 7–27; Norman R. Pace, "The Universal Nature of Biochemistry," *PNAS* 98, no. 3 (2001): 805–808, https://doi.org/10.1073/pnas.98.3.805; L. J. Rothschild, "The Evolution of Photosynthesis... Again?" *Philosophical Transactions of the Royal Society B: Biological Sciences* 363, no. 1504 (August 27, 2008): 2787–2801, https://doi.org/10.1098/rstb.2008.0056; K. W. Plaxco and M. Gross, *Astrobiology: A Brief Introduction* (Baltimore, MD: Johns Hopkins University Press 2011), 13.

5. Martin Chaplin, "Explanation of the Thermodynamic Anomalies of Water (T1–T11)," *Water Structure and Science*, October 14, 2016 (accessed March 16, 2022), https://water.lsbu. ac.uk/water/thermodynamic_anomalies.html#vap. Chaplin writes: "Water has the highest heat of vaporization per gram of any molecular liquid (2257 J x g^{-1} at boiling point).... There is still considerable hydrogen bonding (~75%) in water at 100°C. As effectively all these bonds need to be broken (very few indeed remaining in the gas phase), there is a great deal of energy required to convert the water to gas, where the water molecules are effectively separated. The increased hydrogen bonding at lower temperatures causes higher heats of vaporization (for example, 44.8 kJ mol^{-1}, at 0°C)."

6. Henderson, *Fitness*, 5–6.

7. For recent examples see Stephen Jay Gould, *The Structure of Evolutionary Theory* (Cambridge MA: Belknap Press of Harvard University Press, 2002); and Richard Dawkins, *The Blind Watchmaker* (London: Penguin, 2006).

8. Frederick Copleston, *A History of Philosophy*, vol. 3 (New York: Double Day, 1993), 243.

9. Copleston, *A History of Philosophy*, vol. 3, 127.

10. Stephen Jay Gould, *Wonderful Life: The Burgess Shale and the Nature of History* (New York: Norton, 1990), 291.

11. The idea of man as microcosm is developed in Plato's *Timeaus* and other classical authors. Proclus, a late classical Platonist, believed that "the human animal is a microcosm wherein all elements and all causes of the great universe are found." The classical scholar Plutarch held a similar view, as did Solomon Ibn Gabirol. See the entries under each of these thinkers' names at the Stanford Encyclopedia of Philosophy, https://plato.stanford.edu/index. html.

12. A. J. Gurevich, *Categories of Medieval Culture* (London: Routledge and Kegan Paul, 1985), 59.

ACKNOWLEDGMENTS

I AM GRATEFUL TO SEVERAL SCHOLARS, INCLUDING LAWRENCE HENderson, George Wald, and Harold Morowitz, who defended the fitness paradigm during the twentieth century and whose work provided many of the insights reviewed in this book and in the previous monographs in the Privileged Species series. I am especially grateful to one of France's leading twentieth-century anti-Darwinists, the brilliant polymath Marcel-Paul Schützenberger (Marco to his friends)—a member of the French Academy of Sciences, a colleague of Noam Chomsky, and leading researcher at MIT in the 1950s. It was Marco who first encouraged me to consider the remarkable environmental fitness of nature for life in 1987, while I was on a trip to Paris to attend a genetics conference, by recommending I read Henderson's classic *The Fitness of the Environment*.

I also will always be grateful to Bruce Nichols, my editor at Free Press in New York in the 1990s, who oversaw my first publication in this area, *Nature's Destiny*, and to David Berlinski, who recommended I send the manuscript to his agent in New York, who subsequently sent it on to Free Press.

I am particularly grateful also to John West, vice president of the Discovery Institute, who originally suggested that the evidence of nature's fitness for humankind be reviewed in a series of books—the Privileged Species series. He has been supportive throughout the past seven years while the series was being prepared. I also am very grateful to the many fellows of the Discovery Institute who read preliminary drafts of this monograph and previous monographs of the Privileged Species series and made many critical and useful comments which greatly improved the text.

Last but not least, I would like to pay special thanks to Jonathan Witt. Without his dedication, his attention to detail, his editing skills, and tireless energy, *The Miracle of Man* would never have been published.

FIGURE CREDITS

Figure 1.1. Medieval clock face. "Venice Clocktower in the Piazza San Marco (torre dell' orologio) Clockface." Photograph by Peter J. B. Green, 2006, Wikimedia Commons. CC-BY-SA 3.0. license.

Figure 1.2. A depiction of blood vessels in the human body, from *De Fabrica* by Andreas Vesalius, and a heliocentric model of our solar system, from *De Revolutionibus* by Nicolaus Copernicus, both published in 1543. Public domain.

Figure 2.1. Water in three forms. "Iceberg, Iceland, Ice." Photograph by Sebastian Schellbach-Kragh, Pixabay. Simpfied Pixabay license.

Figure 2.2. The water cycle. Image by John M. Evans and Howard Perlman, United States Geological Survey, April 27, 2005. Public domain.

Figure 3.1. Wildfire in California. "The Rim Fire in the Stanislaus National Forest," Photograph by US Department of Agriculture, August 17, 2013. Public domain.

Figure 3.2. "Solar Spectrum." Image by Robert A. Rohde, October 9, 2008, Wikimedia Commons. CC-BY-SA 3.0 license.

Figure 3.3. The narrow windows in the EM that facilitate photosynthesis. By Brian Gage and Michael Denton.

Figure 4.1. "Purple-Throated Carib Hummingbird Feeding at a Flower." Photograph by Charles J. Sharp, December 1, 2010, Wikimedia Commons. CC-BY-SA 3.0 license.

Figure 4.2. "Mount Everest Morning." Photograph by Ralf Kayser, March 30, 2012, Wikimedia Commons. CC-BY-SA 2.0 license.

Figure 4.3. "Edmund Hillary and Tenzing Norgay." Photograph by Jamling Tenzing Norgay, Wikimedia Commons. CC-BY-SA 3.0 license.

Figure 5.1. The Human Respiratory System. "Poumons2." October 16, 2009, Wikimedia Commons. Public domain.

Figure 6.1. Blood vessels, heart, and lungs. From Andreas Vesalius's *De Humani Corporis Fabrica* (1543). Public domain.

Figure 6.2. "A Red Blood Cell in a Capillary, Pancreatic Tissue." Photograph by Louisa Howard, October 6, 2006, Wikimedia Commons. Public domain.

Figure 7.1. "Polar Bear on Ice in the Southern Beaufort Sea." Photograph by Mike Lockhart, April 8, 2011, United States Geological Survey. Public domain.

Figure 7.2. The Permian reptile Dimetrodon. Painting by Charles R. Knight, 1897, Wikimedia Commons. Public domain.

Figure 8.1. Grains of sand under an electron microscope. Photograph by NASA, no date. Public domain.

Figure 8.2. The pH scale. Arrow pointing to the pH of the bicarbonate buffer. "pH Scale." Image by Christinelmiller, July 3, 2018, Wikimedia Commons. CC-BY-SA 4.0 license. Arrow indicating bicarbonate buffer added by Amanda Witt.

Figure 9.1. "Leaf cutter ants." Photograph by Adrian Pingstone, March 2008, Wikimedia Commons. Public domain.

Figure 9.2. "Organization of Muscle Fiber." Image by OpenStax, 2006, Wikimedia Commons. CC-BY-SA 4.0 license.

Figure 9.3. Bones of the left hand. Illustration by Henry Vandyke Carter, *Gray's Anatomy of the Human Body* (1918), plate 219. Public domain.

Figure 9.4. The human eye. Image by Rhcastilhos and Jmarchn, January 25, 2007, Wikimedia Commons, CC-BY-SA 3.0 license.

Figure 9.5. A computer-generated Airy disc. "Airy Pattern." Image by Sakurambo, July 8, 2007, Wikimedia Commons. Public domain.

Figure 10.1. The NASA rover Curiosity on the surface of Mars. Photograph by NASA, October 7, 2015. Public domain.

Figure 10.2. Blast furnace light in the town of Coalbrookdale, England. Oil painting by Philip James de Loutherbourg, 1801, Wikimedia Commons. Public domain.

Figure 11.1. Two men on the Kalahari starting a fire with a fire drill. Photograph by Ian Sewell, 2005, Wikimedia Commons. CC-BY-SA 2.5 license.

Figure 11.2. The human arm and hand. Illustration by Henry Vandyke Carter, *Gray's Anatomy of the Human Body* (1918), plate 2131. Public domain.

INDEX